With *Burt Rutan's Race to Space*, Dan Linehan tells the dramatic story of Burt Rutan's pioneering aviation work that has included building a racing biplane, the X Prize–winning SpaceShipOne and Voyager, the first airplane to fly around the world.

Linehan gives Rutan the credit he is due as one of the architects of twenty-first century private space travel. As he did with his earlier book, *SpaceShipOne: An Illustrated History*, Linehan also shows himself to be an engaging writer who combines scientific know-how with behind-the-scenes reporting that makes this book read like an adventure story.

—**Paul G. Allen**, *co-winner of the Ansari X Prize*

Dan has done a fabulous job of describing the incredible journey of one of the most accomplished aircraft designers of all time, Burt Rutan. If you weren't impressed by Burt before now, you certainly will be after reading this absolutely fascinating story of the incredible journey of Burt Rutan—from a young model airplane champion to legendary aircraft designer among the ranks of Douglas, Heinemann, Lockheed, and Kelly Johnson.

I personally read it from one end to the other and loved it. This is a book you will read from cover to cover without being able to put it down. What a fascinating story of *the* aircraft designer of our time, Burt Rutan. His accomplishments as an aircraft designer and builder revolutionized the way airplanes are made.

Way to go Dan Linehan for creating a mesmerizing collection of stories!

—**Robert "Hoot" Gibson**, *Space Shuttle Commander*

BURT RUTAN'S
RACE TO SPACE

THE MAGICIAN OF MOJAVE AND HIS FLYING INNOVATIONS

DAN LINEHAN

ZENITH PRESS

First published in 2011 by Zenith Press, an imprint of MBI Publishing Company, 400 1st Avenue North, Suite 300, Minneapolis, MN 55401 USA.

Zenith Press titles are also available at discounts in bulk quantity for industrial or sales-promotional use. For details write to Special Sales Manager at MBI Publishing Company, 400 1st Avenue North, Minneapolis, MN 55401 USA.

To find out more about our books, join us online at www.zenithpress.com.

Designer: Simon Larkin
Design Manager: Brenda C. Canales

Front cover: Top photo by Robyn Beck/AFP/Getty Images; bottom photo by James A. Sugar/*National Geographic*/Getty Images
Back cover: Painting by Stan Stokes

Library of Congress Cataloging-in-Publication Data

Linehan, Dan, 1969-
 Burt Rutan's race to space : the Magician of Mojave and his flying innovations / Dan Linehan.
      p. cm.
 Includes index.
 ISBN-13: 978-0-7603-3815-5
 ISBN-10: 0-7603-3815-9
 1. Rutan, Burt. 2. Aeronautical engineers--United States--Biography. 3. Private planes--United States--Design and construction . 4. Space vehicles--United States. I. Title.
 TL540.R875L56 2011
 629.130092--dc22
 [B]
                                          2010048870

Printed in China

In our modern world, everywhere we look,
we see the influence science has on our daily lives.

Discoveries that were miracles a few short years ago
are accepted of as commonplace today.

Many of the things that seem impossible now
will become realities tomorrow.

—**Walt Disney**, *Man in Space*

In developing aviation, in making it a form of commerce,
in replacing the wild freedom of danger
with the civilized bonds of safety,
must we give up this miracle of air?

Will men fly through the sky in the future
without seeing what I have seen,
without feeling what I have felt?

Is that true of all things we call human progress—
do the gods retire as commerce and science advance?

—**Charles Lindbergh**, *The Spirit of St. Louis*

Contents

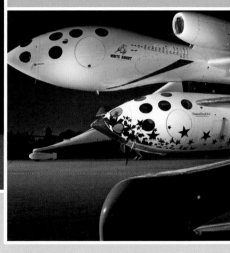

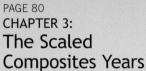

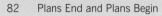

Author Dan Linehan gets the hot seat during this interview (and others) with Burt Rutan in his office at Scaled Composites. *Dan Linehan*

Preface

To cover every detail about every vehicle Burt Rutan had a hand in designing in one book would be simply impossible. As of 2010, a total of forty-four different manned vehicles have been flight-tested between the two companies he founded, Rutan Aircraft Factory (RAF) and Scaled Composites. This figure does not even include unmanned and nonflying vehicles.

This book is not meant to be a biography or a catalog of airplane and vehicle data. It explores how Burt evolved as one of the most remarkable aerospace designers of all time. How did he go from a boy carving his first model airplanes out of balsa wood to designing a spacecraft that opened a whole new realm of space travel? The short answer is that he worked very hard and kept his mind open to innovation. And he was guided by the view that some things may sound impossible, but really they just haven't been made possible yet. Burt was not afraid to try to make things possible.

The first two chapters of this book discuss Rutan's early career and the aircraft he developed for RAF. This period marked the greatest leaps in growth as an aerospace designer for him. Chapter 3 deals with a transition point for Burt as RAF closed its hangar doors for good and Scaled Composites closed its hangar doors to the public. It was during this period that his technological achievements made the greatest gains. However, it is simply not possible to cover all the designs by Scaled Composites in detail for two reasons. First, the large number of projects done by Scaled Composites would require many books to adequately cover them, not just part of a single book. And second, Scaled Composites has a policy of being very tight-lipped about the firm's customers and the details of its projects.

The design of SpaceShipOne, a true example of elegance in engineering, is analyzed in chapter 4. The final chapter culminates with SpaceShipTwo, the vehicle that will realize Burt's childhood dream of someday reaching space.

I was born halfway between the time when humans first orbited the Moon and the time when humans first landed on the Moon. I grew up in an aviation household, with my dad flying for the navy and then for United Airlines. And I was a kid when Burt's designs first hit the air. A combination of these and maybe a few other factors really led me to identify with Burt. I know I am not alone in this.

I had the amazing fortune to work closely with Burt on this book. I was lucky to let my inner child loose a bit. I can't thank him enough for all his time and patience. Thank you, Tonya, for your hospitality. It also was wonderful getting to know Brian, Mike, Dan, and the rest of the Scaled Composites crew over the course of two books, and I very much appreciate the time they could give me.

I am grateful for the assistance I received from great people while writing this book, including those from Virgin Galactic, Vulcan, the Experimental Aircraft Association (EAA), the X Prize Foundation, NASA, the Monterey Navy Flying Club, Mojave Air & Space Port, and many more. This book wouldn't be possible if it weren't for the enthusiasm and support given my first book. I'd like to specially thank Ray at the National Space Society and Kristin at the EAA. Aarzoo, you live up to your name, and I can't express my thankfulness enough to you. And to my family and friends, your understanding kept me going.

—Dan Linehan

Foreword

I am honored indeed to have been asked to pen this foreword for Dan's book. This book is an accurate history of Burt Rutan and his brilliant career. When you, the reader, have read these words and enjoyed the more than two hundred photos included in it, you will have a very complete knowledge of Burt's almost unbelievable technical capabilities and of his fanatical sense of how technology can be used to change and improve aerospace, including some things that weren't meant to fly at all.

I first met Burt at the Experimental Aircraft Association (EAA) convention in Oshkosh in 1974. I was looking for a suitable aircraft design to build in my living room. Burt was selling plans for his first homebuilt design out of the back seat of his new VariViggen, right on the flightline! As it turned out, I was the first "homebuilder" to complete a copy of Burt's VariViggen. My friends and I were a bit concerned about making the first flight of this obviously very different airplane, so I called Burt to get some advice. He told me that he would happily check me out in his VariViggen if I would visit Mojave.

I took my family to California, and we visited Burt in his brand new facility, Building 13 on the flightline of Mojave Airport, home of the Rutan Aircraft Factory (RAF). He seated me in the rear cockpit of his plane and proceeded to demonstrate its amazing flight capabilities as well as its shortcomings. Then he allowed me to fly his creation from the front seat. What a thrill that was and it helped immensely to make for a much safer first flight of my own VariViggen not long after we returned from California.

A year later, my wife Sally (also a licensed pilot) and I flew our VariViggen to Portland, Oregon, and then on to San Jose, California, on a business trip. On the way back we landed at Mojave Airport and visited with Burt. He jumped into my plane and flew it for thirty minutes or so, evaluating it and making several landings. He proclaimed that it passed muster and took us out to lunch. After eating a delicious meal, he offered me a job. On the way home I was so excited I could hardly fly straight! Sally was also very supportive of the idea, and the rest is history, so to speak.

On 21 June 2004, Mike Melvill flew SpaceShipOne to space for the first time. This milestone accomplishment was one of Burt Rutan's biggest personal goals even though this spaceflight did not meet the qualifications to count as an attempt for the Ansari X Prize. It was still the first ever privately funded, designed, built, and operated manned spacecraft to reach space. Burt Rutan and Paul Allen were right there to congratulate the new astronaut as he stepped from SpaceShipOne. *Tyson V. Rininger*

Sally and I joined RAF as full-time employees on 22 September 1978. I was to support those builders who were working on a VariViggen, while Burt would concentrate on supporting the builders of his latest design, the VariEze. Sally became the receptionist/bookkeeper. This state of affairs continued for six months, and then Burt announced that he wanted me to take over VariEze builder support. In order to qualify to do this, he asked me to build a VariEze. I drove to Aircraft Spruce and purchased an entire VariEze raw material kit, including a pair of scissors and a decimal tape measure. I also picked up all available prefabricated parts from Ken Brock Manufacturing (machined parts and weldments) and from Fred Jiran, who fabricated the landing gear, cowling, wheel fairings, etc., at Mojave Airport.

I started building and learning the entire process. When I had completed the canard and elevators for the VariEze, Burt announced that the last thing RAF needed was another VariEze and that he was designing a new flying machine, later known as the Long-EZ. This move was based on feedback received from VariEze builders. The new aircraft would be a bigger version of the VariEze, with lots more range and the ability to carry a starter and alternator. Since I had already completed the canard assembly, he decided to use it on the new design. This is why the VariEze and Long-EZ have the same canard!

For his brother Dick, Burt designed the Voyager, an all carbon fiber design that was intended to fly all the way around the world at the equator without refueling. No aircraft had ever flown anywhere near this distance nonrefueled, and many people, including myself, were skeptical. But in December of 1986, Dick together with Jeana Yeager took off from Edwards Air Force Base and, after circling the globe completely, landed back at Edwards a bit more than nine days later to smash all existing long-distance flight records. The Voyager now hangs in the National Air & Space Museum in Washington, D.C.

Burt is an incredibly prolific designer of composite aircraft. To date, he has fabricated and flown forty-four of his sometimes astonishing designs in the past thirty-nine years—more than one new aircraft design per year! It's a feat no other designer has ever even approached. He is able to literally "see" where the air will go as it flows around the shape he has created. He also has an amazing ability to visualize the loading of a composite structure while in flight. All of his flown designs had good flying qualities and were generally much more efficient than standard category aircraft. More than two thousand of his homebuilt designs have been completed and flown by as many builders. Indeed some are still under construction, and I continue to hear about new first flights several times each year and have done so for the past twenty-five years.

RAF closed down in 1985, and Burt has concentrated his efforts on Scaled Composites, a company he founded in 1982 in order to create a more favorable environment to further his design and engineering skills. He has been one of my best friends for the past thirty-two years. He has been a fair and a generous boss to Sally and me and to all of his employees over the years. Scaled started in 1982 with only six employees, and today has four hundred employees. The work that is done there is state of the art and in some cases has literally been history making.

Burt designed and indeed even physically worked on what was called inhouse the Tier One program. This program became world renowned as we made the first ever successful flights to space that were not paid for by government funds. Prior to SpaceShipOne's first flight to space on 21 June 2004, all flights to space were funded by large governments: Russia, China, and the United States. This achievement was further recognized when three months later the Scaled team flew two space flights within five days of each other, thus winning the $10 million Ansari X Prize for our sponsor, Paul Allen. Paul very generously shared this prize money with Scaled Composites, and the company, in turn, shared the bounty with each Scaled employee, an excellent way to retain great talent when living in so desolate a place as Mojave! SpaceShipOne is now on display at the National Air and Space Museum, hanging above the museum's main entrance between Charles Lindbergh's *Spirit of Saint Louis* and Chuck Yeager's supersonic Bell X-1, with the Voyager not far from sight.

Currently, the Scaled Composites team is busy flight testing a new design known as SpaceShipTwo, a follow-up project funded by Richard Branson's Virgin Galactic and intended to eventually provide rides to space on a commercial basis. Burt recently announced that he will retire in April 2011. It will be interesting to watch the continuing progress of Scaled Composites where Burt has left an unbelievable legacy of truly astonishing aircraft designs and ensured that there is a cadre of exceptional designers, engineers, and test pilots with an unmatched shop full of the best composite fabricators in the world.

—*Mike Melvill*
First commercial astronaut
Tehachapi, California
1 November 2010

Painting by Stan Stokes

Burt Rutan, from Boyhood Dreamer to Aerospace Visionary

"Fortunately, we need not rely solely on governments for expanding humanity's presence beyond the Earth."—Arthur C. Clarke, from the foreword of *SpaceShipOne: An Illustrated History*

The National Air and Space Museum (NASM) of the Smithsonian Institution displays Burt Rutan's most celebrated achievements, including SpaceShipOne, which won the coveted $10 million Ansari X Prize for private spaceflight; Voyager, which hangs with SpaceShipOne in the Milestones of Flight gallery; the Virgin Atlantic GlobalFlyer; and the prototype VariEze homebuilt. His many aerospace innovations preceding his initial designs of SpaceShipTwo and WhiteKnightTwo chronicle a progressive, step-by-step attempt to break barriers with engineering know-how and a wondrous imagination, all the while remaining on the forefront of the burgeoning private spaceflight industry. Rutan's X Prize triumph and his hand in subsequent spacecraft designs are not a beginning, nor an end, but are steps in Burt Rutan's continuing adventure to expand humanity's presence beyond Earth and into space.

Around 1946 to 1957, some mighty big jumps happened in the aerospace community. The jet age began, and so did the missile age. Propellers and pistons abdicated to jet engines and rocket engines as the new state of the art in aerospace. Rutan was between the ages of three and fourteen during these years. He now travels comfortably in the circles of "astropreneurs"—those wealthy individuals who

From 1967 to 1975, Burt Rutan worked only on his own designs for RAF. But as others got to see his work in action, especially how quickly brand-new designs were coming out of the hangar onto the flightline, his drawing board would be used to not just design his own aircraft but also to help design aircraft for customers. *Courtesy of Burt Rutan*

invest sizable amounts of their collective fortune in commercial, manned space programs, people like Paul Allen (Mojave Aerospace Ventures), Richard Branson (Virgin Galactic), Robert Bigelow (Bigelow Aerospace), Elon Musk (SpaceX), Jeff Bezos (Blue Origin), and John Carmack (Armadillo Aerospace).

In terms of their formative years, these astropreneurs grew up during the Apollo space program. Rutan sees them as being young and impressionable during this time period, similar to children who grew up in the early years of flight and went on to become leaders in twentieth-century aviation.

To commemorate the one-hundredth anniversary of the Wright brothers' first flight in 1903, *Aviation Week* asked Burt Rutan to come up with a list of the ten people he thought had been the most influential in aerospace since this milestone. His list included Wernher von Braun, Kelly Johnson, Charles Lindbergh, and Howard Hughes, among others.

"Everybody I put on my list was a young child during this big spurt of aviation progress from 1908 until the First World War," Rutan said. Before this period, less than a dozen people had ever piloted an airplane. Aside from the Wrights, most flying had been in a straight line. That all changed in 1908 when aviation learned

to veer into more adventurous paths and the aviation renaissance really bloomed.

Like Rutan and other witnesses to the Apollo mission milestones, the pioneers in the early decades of flight were impressionable kids when this new invention sparked their imaginations. "You can look at anything that's high technology," Rutan said. "It happens in spurts. Had we not had Apollo, for example, if we had beaten the Russians with Alan Shepard instead of Yuri Gagarin, or, if we had beaten the Russians with Explorer instead of Sputnik—those two races were only weeks apart, not years or decades—and had we not found the need for national prestige to do something great like go to the Moon, it is possible—and this is just a gut feeling from me, there's very little proof—that these billionaires who are fascinated and enthralled by commercial space may have spent their money on other things if they weren't, as children, shown that enormous progress can be done in a very short period of time."

Burt Rutan has two dominant parts to his personality; he can be focused, and he can be severe. But he also has a good wit to temper his outspoken nature. For instance, he is not shy in his criticism of NASA, especially regarding what he sees as its retreat from the ambitious goals of the Apollo generation, the kind of goals that led

A flip of the coin determined who would fly the Wright Flyer in 1903 from Kill Devil Hill outside Kitty Hawk, North Carolina. Orville won the toss. His flight covered 120 feet and lasted 12 seconds. It was the first time a powered aircraft, heavier than air, flew under control and over a sustained period of time. *NASA*

A wonderful comparison between the capsules and rockets of NASA's early spaceflight missions, the diagram shows the single-person Mercury capsule, the two-person Gemini capsule, and the three-person Apollo capsule, which is sandwiched between its rocket-powered service module and gangly lunar lander. A much larger difference exists between the rockets used to get these capsules into space, Apollo's Saturn V (left), Gemini's Titan (center), and Mercury's Atlas (right). *NASA*

Rutan to dream of outposts on Mars and other planets.

Rutan once said, "The thing you got to do is to always challenge yourself with something that you don't think you can do. . . . If you think you can do it, and you maybe even know that you can do it, it's not true research because you can't be innovative. You can't have a breakthrough. You're removing your opportunity to do that."

In his view, NASA's struggles in moving manned spaceflight out of Earth's orbit are directly related to its culture of thinking small over the past few decades. "All the stuff they're doing they know will work, which means they can't stumble into new ideas to help us get to more interesting places than the Moon—the moons of the outer planets that actually have oceans below the surface. And we are not going to get there unless somebody goes out and takes risks and tries stuff that may not work."

Today, a simple, ordinary-looking manila folder of model numbers exists in Burt Rutan's office. Once opened, pistons roar, jet wash spills out, and rocket plumes light up the room, like some kind of aviation mysticism.

It is thick with page after page of flying machines—record breakers, racers, world flyers, spaceships, and many other aerial innovations. Some eventually spring to life. Some don't.

"Anytime I get something that is at least to a three-view and a spec and something—even if it is not necessarily going to be built—I put a model number on it," Rutan said.

With a grin that looks as if he's about to show off a map to secret treasure or a book of magic spells, he says, "I'll give you a peek at it, so you can see the kind of depth of it. It is not very deep."

It didn't take much to spark a gathering. This particular event drew nearly ninety Rutan designs and spinoffs. Taken at a time when Scaled Composites was in full swing, this photograph cannot show the other good things inside the hangar. *Courtesy of Burt Rutan*

Oh, but it certainly is. The file includes a napkin sketch for Voyager he made for his brother, Dick Rutan, and Jeana Yeager. Dick and Jeana would go on to pilot Voyager around the world.

His first manned aircraft to fly, the VariViggen, is Model 27. He was able to go back and distinguish twenty-six prior versions that directly led to this aircraft. They were model airplanes, and they were also concepts that never left the paper. But each one was a step toward flight.

"I just assigned a new model number this morning, 357," he said in June 2009. "And that's—I can't tell you what it is. But, 356 was a turboprop Boomerang, which we are doing for Dale Johnson, who wants to put the Boomerang in production as an eight-place turboprop. Dynamite airplane."

In many cases, the model numbers ended up being painted on airplanes as tail numbers. For instance, the tail number of Virgin Galactic's prototype SpaceShipTwo, Model 339, is N339SS. All aircraft from North America have tail numbers that begin with *N*, and *SS* in this case stands for *SpaceShip*. The VariViggen prototype is N27VV, which is straightforward to decode. Sometimes customers have their own tail number in mind. Danny Mortensen's prototype AMSOIL Biplane Racer, Model 68, is N301LS. "Sometimes I just put dashes. It is not a hard and fast thing that I stick to," Rutan added.

Burt Rutan's Race to Space seeks to reveal how these model numbers in Burt's prized folders tranformed into revolutionary aircraft and spaceships. We'll start by focusing on his design work on the F-4 Phantom at Edwards Air Force Base, when he was a flight test project engineer in the years 1965 to 1972, and move onward and upward to SpaceShipOne and SpaceShipTwo, the designs that announced the birth of the private spaceflight industry to the world.

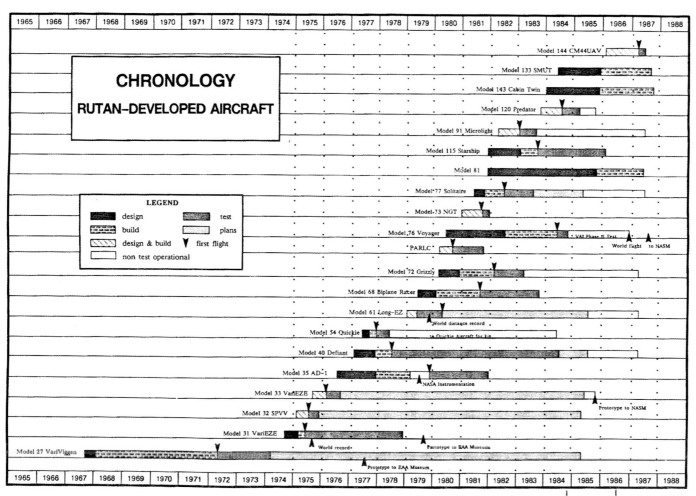

This diagram, first published in *Canard Pusher,* shows the project timelines for Burt Rutan's first twenty-one designs. The first fifteen were Rutan Aircraft Factory vehicles, which included the U.S. Navy Power-Augmented Ram Landing Craft (PARLC). The Boomerang design began several years after this time period. The last six were part of Scaled Composites, a new company founded in 1982 to work on proprietary stuff. *Courtesy of Burt Rutan*

Burt Rutan wasn't the first to use canards, but he certainly was the first to make an art form out of his planforms. The VariViggen, with its hammerhead-shaped nose, sweptback fighter-like wings, and external fuel tank, was the first airplane that he designed, built, and flew. *Courtesy of Burt Rutan*

Chapter 1

Rutan Lifts off the Ground

Burt Rutan's work on the F-4 Phantom proved to be a substantial influence not only on the first manned aircraft he designed and flew but also on many others to follow. The F-4 faced a deadly problem with unrecoverable flat spins. The only thing the pilots could do when trapped in this situation was to yank on the ejection handle, if they were lucky enough to have time to do so.

Designed during World War II but first flown after the war, the Convair B-36 Peacemaker was the largest bomber ever built. It proved to be an early inspiration to Burt Rutan. With a wingspan of 230 feet, the B-36A had six Pratt & Whitney R-4360 radial pusher engines. Modifications to later models included the addition of four jet engines. *USAF*

Rutan was able to help develop a procedure where pilots could reduce the probability of entering a flat spin after the aircraft departed from controlled flight, allowing them to return to base with the aircraft intact. In order to accomplish this, Rutan, in the backseat, and test pilot Jerry Gentry flew the fighter through more than a hundred departures—a departure is when the aircraft does not respond to the pilot's control inputs—and normal spins with the aid of a specially designed spin-recovery parachute. They became the first ever to recover from a flat spin in an F-4 Phantom.

In an article Rutan wrote for *Sport Aviation* shortly after leaving the USAF, he stated: "Stall/Spin is the major cause of general aviation fatal accidents. NASA and the FAA are currently investigating the problem by evaluating and correlating the spin characteristics of current designs, evaluating new pilot training procedures, proposing instrumentation, and other methods. I feel strongly that too little emphasis had been placed on designing the overall configuration for safe high angle of attack flying qualities and that not all the important criteria are being considered."

His solution to this was to create an aircraft that would not stall.

Above: **When Burt Rutan** went to the store to buy his first model airplane, at an age too young for him to remember, his mother worried that he would become frustrated by not being able to put together the prefabricated parts. To her surprise, he returned with only blocks of wood and then carved his own airplane. Years later, he would go on to become a champion model airplane builder and flyer.
Courtesy of Burt Rutan

Right: **Walt Disney** (left) and Wernher von Braun (right) show off a model of the XR-1 that was featured in a three-part Disneyland television program that Burt Rutan watched as a kid in the 1950s. The XR-1 was the last stage of a four-stage rocket that von Braun designed to fly astronauts into orbit and then glide back down to Earth. *NASA*

Year	Left Events	Right Events
1965	Flight Test Engineer for USAF, Edwards AFB, CA and St. Louis, MO (6/65-3/72)	Graduated from Cal Poly, San Luis Obispo, CA (1965)
1966		
1967		
1968		Construction of VariViggen begins (1968)
1969		Rutan Aircraft Factory (RAF) founded (1969)
1970		
1971		Bought BD-5 kit to use parts to build MiniViggen (1971)
1972	Director of Bede Test Center, Bede Aircraft, Newton, KS (3/72-5/74)	VariViggen first flight, Newton, KS (5/72) VariViggen first flown to EAA Fly-In Convention, Oshkosh, WI (7/72)
1973		
1974		Rutan relocates to Mojave, CA, to work full-time on RAF (6/74)
1975	VariEze POC first flown to EAA Fly-In Convention, Oshkosh, WI (7/75) VariEze POC breaks distance world's record, Oshkosh, WI (8/75)	VariEze POC first flight, Mojave, CA (5/75) VariViggen SP first flight Mojave, CA (7/75)
1976		Homebuilt VariEze first flight, Mojave, CA (3/76) Homebuilt VariEze first flown to EAA Fly-In Convention, Oshkosh, WI (7/76)

Above: **During the twelve-year period** shown in the timeline, Burt Rutan started his aerospace career as a flight test engineer at Edwards Air Force Base. He worked for Bede Aircraft afterward and then returned to Mojave to work full time on his own company, Rutan Aircraft Factory. Over this time, he made first flights on four models that he designed, one of which set the first of many world records his designs would come to earn. *Dan Linehan*

Right: **Burt Rutan's futuristic design** drew attention from the farthest reaches, even from a galaxy far, far away. Never noticed this before, but Vader also sports a serious set of sideburns. *Courtesy of Burt Rutan*

The Stall

It is extremely important to understand what a stall is and how a stall is caused because conquering the stall was the primary focus of Rutan's early designs.

A stall occurs when air no longer flows smoothly over a wing, causing the wing to lose its ability to create lift. Without lift, airplanes don't fly. They become big, giant falling objects. Stalls are especially dangerous at low altitudes, like when landing or taking off. A pilot needs time to correct the problem, and altitude equals time. Also, when a hazardous spin occurs, it is typically trigged by a stall.

So what happens to the air to cause it not to flow smoothly over the wing in the first place, eventually leading to a stall? A wing has a shape, known as an airfoil, specially designed to interact with the air to cause lift. This favorable interaction only occurs when the wing moves through the air within a range of orientations. Once outside this range, when the airplane's angle of attack exceeds a critical value, the airfoil fails to properly provide the expected lift and the aircraft departs from controlled flight.

But angle of attack is the key term here.

Imagine a pilot flying straight and level. The angle that the aircraft makes with the oncoming air is the angle of attack. Or, more precisely, the angle between the relative wind and the chord line, which is a reference line drawn from the front edge of the wing's airfoil cross-section to the back edge. If the pilot pulls the nose up and the aircraft is still flying straight and level, the angle of attack has increased. The angle that the aircraft's nose points in relation to the direction of the air's movement has increased. So if the pilot pulls back even more while still flying straight and level, the angle of attack becomes even greater. At some point this angle reaches the critical value and the air moving over the wing goes haywire.

Generally speaking, if the pilot pulls the stick back too much, causing the nose to rise too much, and airspeed is too slow, then the condition is ripe for a stall.

As a flight test engineer for the USAF, Burt Rutan (third from the right) performed his duties in the backseat of the McDonnell Douglas F-4 Phantom. A frontline fighter-bomber during the Vietnam War, the F-4 had a problem where under certain conditions it would depart into a nonrecoverable flat spin. Rutan helped devise a solution for pilots that used a drogue chute to avoid entry into the flat spin.
Courtesy of Burt Rutan

Conventional Aircraft

Normal Flight

Airflow

Main wing creates all lift   Chord line   Horizontal stabilizer pushes downward

Slower Flight

Airflow

Angle of attack equals 10°

Very Slow Flight and Stalls

Airflow

Angle of attack greater than stall limit

Separated airflow is highly chaotic and turbulent, causing main wing to stop lifting and stall

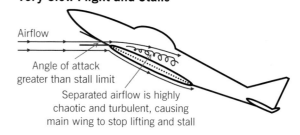

Three views of a conventional aircraft flying straight and level but at different speeds: As speed decreases, the nose pitches up, but the direction of the oncoming airflow stays the same. A stall occurs once the angle of attack becomes too large. At this point, air can no longer flow smoothly over the wing, and the wing can no longer generate lift. *Dan Linehan*

The Canard

So Rutan sought to eliminate the dangers poised by stalls by effectively designing them out of the airplane. He did this with the canard, the little tail attached to the nose instead of the backend of the fuselage like on most other airplanes.

"I was entranced for some reason—I don't know why—by the Saab Viggen," Rutan recalled of the Swedish fighterjet back during his college days. "The B-70 hadn't flown yet. There were pictures of it, this new, super, Mach 3 bomber that's being built."

Both these aircraft had canards. While still in college, Rutan did some canard building of his own.

"I found out with a RC model, and I found out by doing a little wind tunnel model, that I could get natural stall limiting. And I was absolutely fascinated by that," he said. "Whether it was forward or aft CG, I could take the airplane full aft stick and run up to a certain angle of attack and then pull all you want and it doesn't go higher. This is a big safety thing."

As the end of Rutan's USAF career neared, he began to work on the brand new F-15 Eagle.

"I was intimately familiar with the flight control systems on the F-15 because I was an air force guy, a

Below: If there was one aircraft that echoed in the mind of Burt Rutan as a young engineer, it had to be the Swedish Saab 37 Viggen. The multirole fighterjet used a canard to give it better short landing and takeoff ability. With a max speed above Mach 2, the Viggen had an approach speed of only 137 miles per hour. *Saab*

Canard Aircraft

Normal Flight

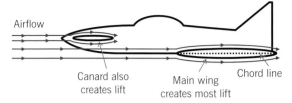

Airflow

Canard also creates lift

Main wing creates most lift

Chord line

Slower Flight

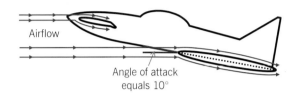

Airflow

Angle of attack equals 10°

Very Slow Flight but Does not Stall

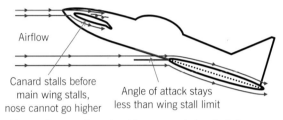

Airflow

Canard stalls before main wing stalls, nose cannot go higher

Angle of attack stays less than wing stall limit

Above: For an aircraft with a canard that is flying straight and level, as it slows down, the nose also pitches up like a conventional aircraft. However, as the angle of attack increases, the canard is designed to stall before the wing. When the canard stalls, the nose cannot rise any farther. This prevents the wing, and thus the aircraft, from stalling. *Dan Linehan*

flight test planner, getting ready for the air force to flight test the F-15 in a few months."

The F-15 didn't have fly-by-wire, a flight control system using a computer to limit how a pilot can fly a plane, as did the F-16 Fighting Falcon, which entered service afterwards. The F-15 didn't have a canard either, so it did not have natural stall limiting.

"I was fascinated by having no electronics, no servos, no control augmentation, no sensors, no hydraulic system, and still getting natural stall limiting," Rutan said.

There is another advantage to using a canard. The horizontal stabilizer of the tail for a conventional aircraft is a small wing just like the canard is a small wing. Besides the obvious that the horizontal stabilizer is in the back of a conventional aircraft and a canard is in the front, there's a very big difference in how they operate. The horizontal stabilizer is actually designed to push with a downward force. In other words, a horizontal stabilizer doesn't create lift, it does the opposite. However, the canard doesn't behave this way. It actually creates a lift.

So in a conventional aircraft the main wing creates the lift and the tail's horizontal stabilizer doesn't. This makes the main wing have to lift more. But for an aircraft with a canard, both the canard and the main wing create lift.

How exactly does stall limiting work with a canard?

The canard prevents an aircraft from stalling by preventing the main wing from stalling. The canard does this by actually stalling itself at an angle of attack lower than the angle of attack that will trigger a stall on the main wing. If the pilot pulls back too much on the stick, all of the sudden the lift normally created by the canard does not increase any more. As a result, the nose does not go any higher, even when the pilot pulls the stick back further. So the pilot cannot force the wing above its stall angle of attack.

And Rutan took full advantage of this.

Burt Rutan created is his own wind tunnel, and as long as the pavement didn't run out, there was plenty of wind to test his designs. By attaching a model to an array of sensors mounted on top of the family car, he was able to collect flight data from a set of instruments as he sat in the passenger seat. *Courtesy of Burt Rutan*

Above: An encounter on a runway with a North American B-70 Valkyrie and its row of afterburning jet engines more than resonated with Burt Rutan. Capable of flying Mach 3.1, the B-70 had a canard to control pitch and wingtips that folded downward to improve stability during supersonic flight. Only two were ever constructed. B-70 test pilot Fitz Fulton would eventually fly for Scaled Composites. *NASA*

Left: Burt Rutan spent several months in St. Louis at McDonnell Douglas working on the brand new F-15 Eagle. It flew at a top speed of Mach 2.5 but did not have a fly-by-wire flight control system as most modern fighters now do. Rutan left before the F-15 began flight testing to work for Jim Bede. *USAF*

Painting by Stan Stokes

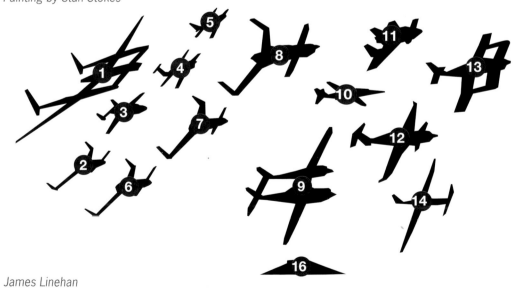

James Linehan

Designs from Rutan Aircraft Factory, 1972 to 1996

1—Voyager (Model 76)
2—VariEze POC (Model 31)
3—Next Generation Trainer (Model 73)
4—AMSOIL Biplane Racer (Model 68)
5—Quickie (Model 54)
6—VariEze Homebuilt (Model 33)
7—Long-EZ (Model 61)
8—Defiant (Model 40)
9—Boomerang (Model 202)
10—AD-1 (Model 35)
11—VariViggen (Model 27)
12—Catbird (Model 81)
13—Grizzly (Model 72)
14—Solitaire (Model 77)
15—PARLC
16—Mojave Pyramid House

VariViggen

It is not coincidence that Burt Rutan drew inspiration for the VariViggen from military aircraft. The construction of the VariViggen began while he still worked as a flight test engineer for the USAF. Now that the sleek, giant, and futuristic-looking B-70 Valkyrie was flying, Rutan got to see it up close, very close.

"It's sitting at the runway for takeoff at Edwards," Rutan said of the B-70. "We're number two for takeoff, Jerry Gentry and me in an F-4 Phantom. We have the canopies up, in very close proximity to the Valkyrie. We are not behind, of course. We are just a little bit to the side. It's holding its brakes, and it runs up to full military power. Then, one at a time, we see all six of those engines go into afterburner. The heat and noise were so fierce that it hurt, even wearing our helmets. We closed the F-4's canopies to avoid serious ear damage. The Valkyrie starts rolling, rolling. After it starts moving, they clear us into position and hold. So we are sitting on the runway now, watching it go down the runway directly in front of us."

Moving his hand over an imaginary three-mile-long runway and then lifting off, Rutan continued, "It comes up like this—nose way above its tail—and you get to see the full planform of it as it staggers into the air. I'll never, ever forget that sight. That was very cool."

While writing this book, I found myself running along quite a distance of runway fence at Mojave, two days after the unveiling of SpaceShipTwo, trying to get a better look at Proteus doing touch-and-go landings. I wonder how much Rutan realizes that decades later, people like me get that same type of feeling seeing one of Scaled Composite's designs lifting off into the sky.

VariViggen Design

The other major inspiration for his first aircraft was the Swedish Saab Viggen. This supersonic fighter sported a canard, a small wing forward of the main wing, that gave it excellent short takeoff and landing capabilities. Besides, it was really a great looking airplane.

"In those days, I was a back-seater only, of course," Rutan admitted. "I always was. I wasn't a rated military pilot. I was a flight test engineer. So I got to go Mach 2. I got to pull 8 g. I got to go to fifty thousand feet. But I'm in the back, and the blue-suiter is flying. So I wanted to have my own fighter."

The airplane Rutan wanted for himself had to have a phenomenal roll rate and be able to turn like a fighter. It had to be very fun to fly.

"I had done these very rough designs and wind tunnel tests for it while I was still in college," Rutan said. "And out of college I decided that I needed a lot better data to evaluate its feasibility. I had sorted out an interesting way to get trimmed, three-axis stability and control data on a new type of wind tunnel model mount. I used the idea on a rig on top of a car, driven down the road to make the needed wind. The setup had a little flat plate, so it measured the dynamic pressure, so I could also measure model drag."

He built an instrument panel with voltmeters on it, and he had a little reel-to-reel tape recorder. With somebody else driving the car, he'd sit there changing the elevator position, which would change the angle of attack.

"I read all these instruments into the tape, and then later I'd listened to the tape and write everything down. I'd thus have flight test data I needed. This was done

Construction of the VariViggen began in 1968 while Burt Rutan worked at Edwards Air Force Base. However, the ideas he had for its design began when he was an engineering student at California Polytechnic State University in San Luis Obispo. *Courtesy of Burt Rutan*

The A-12, "tail number" FX-935, was Burt Rutan's official Model 1. The Saab Viggen strongly influenced the design of this RC model. Featuring a canard and push-pull engines, the A-12 evolved twenty-six designs later into Model 27, the VariViggen. *Courtesy of Burt Rutan*

while my day job was flight test engineer on the XC-142, the F-4, and other projects."

By using his car top wind tunnel, he now felt comfortable that he indeed had a design with natural stall limiting that would also be very fun to fly. Rutan didn't focus on optimizing the airplane's speed, range, or efficiency.

"You know it wasn't a very good airplane for performance, particularly up high because it had a lot of induced drag," Rutan said.

What is considered not very good performance is relative, especially when goals like building the most efficient airplane are considered. Rutan would also be quick to point out that many of the aircraft flying then did not have very good performance.

The VariViggen was a first step. However, he did not want to call the VariViggen his Model 1, even though it was his first full-sized airplane to fly.

"I built that RC model that was a totally different airplane. And I modified it. Then I tried something else in a wind tunnel. I went back and tried to identify every change that was substantial that I had made in developing the VariViggen from a design standpoint. I came up and said this will be Model 27."

What would then become Model 1 didn't look at all like the VariViggen, but it looked like the Saab Viggen. That was about it for similarities, though. A single-place design, the pilot would lie on his stomach. But the model of the tiny airplane, calling for a two-stroke pusher engine, wouldn't even balance.

"It was a dangerous thing. I mean it was crazy," Rutan said.

Rutan had been building airplanes nearly his whole life. In 1968 at that age of twenty-five, now a flight test engineer and already a pilot, he began construction. To this point, his models had all been made out of wood. Using classic airplane building techniques, he also constructed the VariViggen out of wood, with two noteworthy exceptions. To gain experience working with metal, he built the outer wings and the rudders out of aluminum.

"When I built it, I didn't take pictures, and I didn't make drawings of it. I built this airplane without making drawings because, from model airplanes, I knew how to build an airplane. I knew how to make stress calculations. I knew how thick the spar I needed, how big a bolt I needed to hold the wings on. I made those calculations."

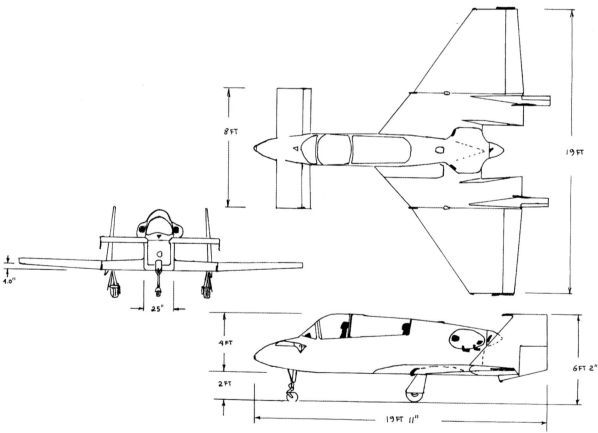

Designed to be Burt Rutan's own fighterlike airplane, the VariViggen, shown here in three-view, looked more like an airplane found at an air force base rather than at a town's local airstrip. The canard, bubble canopy, sweptback wing, twin vertical stabilizers, and pusher engine gave it an undeniable futuristic aurora when the VariViggen first began to fly. *Courtesy of Burt Rutan*

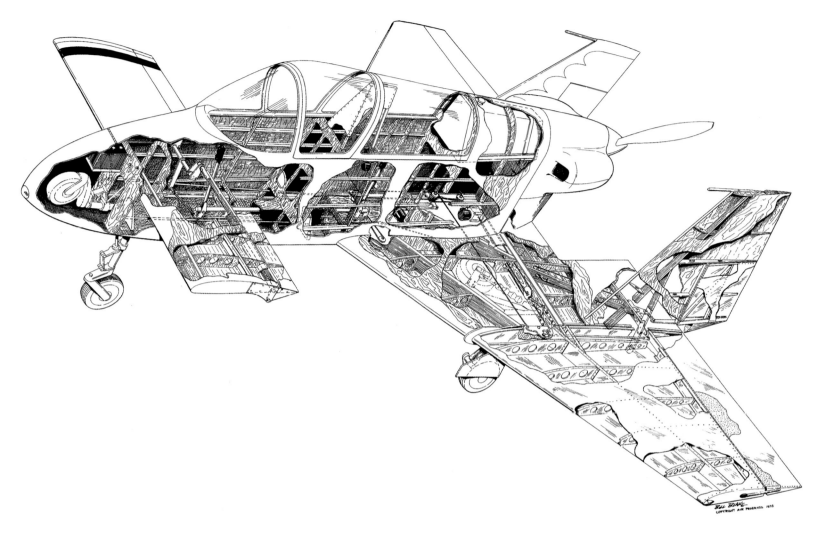

Fun to Fly

Rutan now had himself a jet fighter, just without the jet engine. "I didn't think about selling plans then," Rutan said about the aircraft initially. "The VariViggen was for me."

In 1972, still an employee of Bede, with only seventy-five hours flown on the VariViggen, Rutan flew to the Experimental Aircraft Association's (EAA) annual fly-in and convention held at Oshkosh, Wisconsin. Known as AirVenture, it remains one of the premier air shows in the world. Originally founded to support homebuilders, the EAA now draws all matters of plane, pilot, and aerospace enthusiasts.

This was an impressive stage for the young upstart Rutan to make his first showing. And the crowd went wild for the VariViggen. Reaction was so strong that in less than two years he would strike off on his own and start selling plans.

In the meantime, Rutan didn't miss many opportunities to show off his new creation. He and his second wife, Carolyn, returned for a visit to the

Though Burt Rutan is very well known for his innovative use of composite construction, this cutaway drawing shows how the VariViggen's fuselage was built out of wood and its wing was made with aluminum. *Courtesy of Burt Rutan*

When this photograph of Carolyn standing on the wing of the VariViggen appeared on the cover of *Sport Aviation* in August 1973, it created a huge stir. A lot of VariViggen homebuilt plans sold, but it was the last time *Sport Aviation* ever had a sexy, pinup-style cover. *Courtesy of Burt Rutan*

McDonnell Douglas plant in St. Louis, where they had met while Rutan worked on the F-15 for four months. The headset in the backseat was broken, though. During the flight, Rutan would have to shout back to keep her appraised of the flight progress.

It was a treat to land at St. Louis International. The airport had commercial airliners on the south side of the field. But on the north side of the field, McDonnell Douglas was building jet fighters. They shared the same runway; an F-4 would be taking off with afterburners followed by an airliner en route to some international destination.

"There was a new, big commercial control tower, and when I flew in, they were really blown away by the look of this homebuilt," Rutan said.

The VariViggen just didn't look like any other homebuilt at the time. After landing and taxiing up to McDonnell Douglas, a member of the flight test team wanted a ride. Rutan obliged. They taxied out but had to wait quite a while for airliners.

"I had landed a couple hours before. I think it was the same shift that was there, and they saw this thing fly in. They were really curious about getting a better look at this thing. They called me before they cleared me on the runway, and they said, 'Listen, we'd really like to get a better look at that. Could you come out here and maybe circle the tower when you takeoff.'

"And I said, 'Well, how close do you want me to fly to the tower?' They then said, 'Use your discretion.' Wow, I'm getting that *discretion* freedom from a FAA controller in a control tower at a large commercial airport."

Because the backseat headset was broken, it was hard for Rutan to communicate this to the passenger in the backseat. And the tower had just cleared them for immediate takeoff. Airliners waited behind them.

"He's thinking I'm kind of a wild guy anyway cause I'm out here in Kansas flying homebuilt airplanes," Rutan said. "And I don't work for the air force anymore, so I might not be real strict about the normal rules. He is a flight test pilot who flies F-4s and F-15s in a mixed commercial airport and who knows how important it is to do everything by the book. And he's in the back in the VariViggen.

"We just rotated. I turned like this, and I headed straight for the control tower. I rolled it up, and I jogged a little bit to the right. I went by the control tower with 90 degrees of bank and full aft stick like this," Rutan said, carving a tight circle with his hand. "And you could see the guys in there and tell what kind of glasses they were wearing. They were running around the tower to watch me come around. And it was no big deal because I was cleared to use my discretion. And maybe I wasn't that close, but the story always gets better after more years."

Burt looked back in the cockpit and watched his passenger take off his badge with despair.

"He thought we were in big trouble 'cause Burt Rutan had buzzed the control tower," Rutan said with a wide smile.

Plans and Kits

Plans for the VariViggen were completed during the fall and winter of 1973. By May of 1974, homebuilders bought 190 sets of plans. With Rutan's kitplane design experience from his time at Bede, he decided to take a unique direction for his VariViggen homebuilt. He didn't want to have a big company or get involved with large inventories of parts, especially since he couldn't afford to spend hundreds of thousands of dollars on stocking up the first hundred kits.

"I got with Ken Brock to have him make machine parts and welded parts in his shop," Rutan said. "I

Even the cockpit of the VariViggen had the look and feel of a fighterjet. Stall resistant, high performance, and fun to fly, the VariViggen could be built by homebuilders from a set of plans.
Courtesy of Burt Rutan

met him through EAA. He was doing gyrocopter kits. Fred Jiran Glider Repair over here in Mojave was doing fiberglass sailplanes. I had them build three composite parts for the VariViggen that the homebuilders could buy: the cowling, the nose cone—the pretty part with the nose light in the middle, and this combing that comes up and forms the windshield frame and the instrument panel mount and so on."

These curvy parts were easier to build with fiberglass than wood. Jiran also sold these parts, passing on a small cut. Rutan also made a deal with Aircraft Spruce to put together the VariViggen kit. He, in turn, sent homebuilders their way.

"They are in that business," Rutan said. "So I just gave them a spec showing them how much of this kind of wood and glue and nuts and bolts and instruments and so on."

To help keep costs down for the homebuilders, on his subsequent aircraft, he had Wicks Aircraft Supply also make kits. Both kit supply companies forked over a cut as well.

"RAF got 7 percent without having any risk of excess inventory and without putting a lot of money up front.

So that was one of the smartest things that I did as a businessman. I was sort of in the kit business, but I didn't have to have employees and the warehouse and the catalog. So, I would send people there. They would buy a VariEze kit, and I would get a small cut of it. That's how we survived, even though we sold a set of plans for the VariEze at $128. And only $27 for the VariViggen."

The VariViggen was a challenging aircraft to build. After all, it took Rutan four and a half years to build it himself. Plans sold, but only a few dozen eventually took to the air. Rutan realized that he would make only $27 on a set of plans, plus the small cut from the kits, but would have to support the homebuilder for ten years during the construction, followed by the support needed to fly the completed homebuilt aircraft.

"It's not a real good performing airplane, while it is the most fun thing in the world to fly, pretending to be a fighter pilot. And you can beat anybody on spot landing contests. I decided I'm just not going to sell plans for it anymore."

RAF stopped selling VariViggen plans in 1978, but builder support continued.

VariViggen Details

Model number	27
Type	single-engine, canard pusher
Prototype tail number	N27VV
Current prototype location	EAA AirVenture Museum, Oshkosh, WI
Customer	homebuilders, marketed 1974
Fabrication	RAF
Flight testing	RAF
First flight date	18 May 1972
First flight pilot	Burt Rutan
Seating	two-place, tandem
Wingspan	19 ft
Wing area	119 ft²
Aspect ratio	3.03
Length	19.9 ft
Height	6.2 ft
Empty weight	1,020 lbs
Gross weight	1,700 lbs
Engine	Lycoming O-320-A2A, 150 hp (4 cylinders)
Landing gear	tricycle, retractable
Fuel capacity	30 gal
Takeoff distance	850 ft
Landing distance	300 ft
Rate of climb	800 fpm
Maximum speed	165 mph
Cruise speed	150 mph
Range	300 miles
Ceiling	14,000 ft

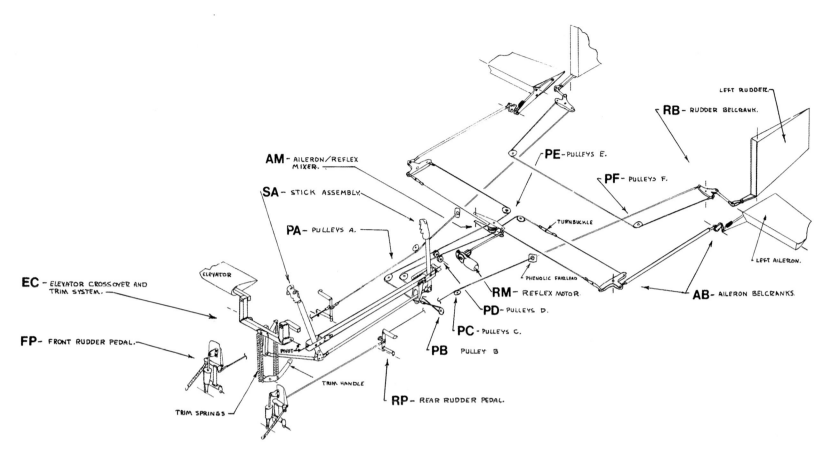

AM - AILERON/REFLEX MIXER.

SA - STICK ASSEMBLY.

PA - PULLEYS A.

EC - ELEVATOR CROSSOVER AND TRIM SYSTEM.

FP - FRONT RUDDER PEDAL.

ELEVATOR

PIVOT

TRIM SPRINGS

TRIM HANDLE

RP - REAR RUDDER PEDAL.

PB PULLEY B

PC - PULLEYS C.

PD - PULLEYS D.

RM - REFLEX MOTOR.

PHENOLIC FAIRLEAD

TURNBUCKLE

PE - PULLEYS E.

PF - PULLEYS F.

RB - RUDDER BELCRANK.

LEFT RUDDER

LEFT AILERON.

AB - AILERON BELCRANKS.

Above: As shown in the diagram, the yoke—or control stick—moves the elevators on the canard for pitch control and the ailerons on the wing for roll control. Rudder pedals move the rudders on vertical stabilizers for yaw control. *Courtesy of Burt Rutan*

Right: The word canard comes from the French word *duck*. In English, it means a *hoax* or something *misleading*. It came to be used for airplanes because as a duck flies, its wings are closer to its tail than head, compared to most other birds. However, the VariViggen was not an ugly duckling and made its film debut in the sci-fi flick *Death Race 2000*, with Burt Rutan as stunt pilot for the movie. *Courtesy of Burt Rutan*

MiniViggen

A year before Jim Bede approached Burt Rutan to work for him, Rutan had bought a BD-5 kit. Made of metal, the small and sporty BD-5 had a single seat, straight wings, a conventional tail, and a pusher engine. It too looked like a mini fighter. Rutan wasn't planning on building the BD-5. He was planning on designing his own canard airplane based on the BD-5. Using the same fuselage but now a two-seater, it would have a sweptback, high-wing configuration with fins at the wingtips.

"Laying the kit out on the floor, hey, it had landing gear, an engine, a driveshaft, a propeller, instruments, and rod ends for flight controls," Rutan said. "I could use a vast majority of the parts of this kit. I would sell a kit myself, which were just the things that they would need to make this MiniViggen."

So Rutan's first idea for an aircraft to sell was actually the MiniViggen, not the VariViggen.

When Rutan did start working for Bede in March of 1972, his VariViggen had not yet flown, and he didn't tell Bede about his idea to make the MiniViggen. The BD-5 kit was one of Bede's own kits after all. It was way too early in the process to add this complication, and the distance between this idea and the realization of it was too great.

Rutan started building parts of the MiniViggen in his basement and garage in Newton, Kansas.

"I was going to take the BD-5 kit and have my builders build a canard airplane," Rutan said. "I had to have new ribs and wing spars and everything. I decided to build a fiberglass elevator for the canard. So my very first thing that was composite work was to build the elevator for the MiniViggen."

Rutan zoomed around in his VariViggen at air shows now that it was flying. At his day job, he worked on the new jet version of the BD-5, the Model J. All this as he continued to tinker with the MiniViggen. But after two years at Bede, he decided it was time to move on.

Change in Course

"I thought I'd use 85 percent of the kit. As I kept trying to develop this idea, the plans for this MiniViggen, man, I was down to 15 to 20 percent of the kit. A guy would have to buy this kit and throw away most of it. That didn't make a whole lot of sense."

Rutan found the construction to be challenging and time consuming as well. The metal structure had other drawbacks, such as heavy weight. Model testing also revealed stability issues.

There had to be an easier way.

"I was kind of bailing on that. I decided that maybe I can make a living selling VariViggen plans and parts. I needed some kind of job now that I was going to quit Bede. I didn't want to go back to the air force. I did want to come back to California."

After borrowing some money from his pop, he hopped in an old, 1946 Ercoupe he borrowed from his uncle and flew around California, scouting for a place where he could afford a house, a shop, and a hangar. Bouncing from airport to airport, he touched down at Mojave. Not only did Mojave meet his needs, Rutan was returning to familiar territory, with Edwards only twenty miles due southeast.

"I needed to have a hangar to put my VariViggen in," Rutan said. "I needed to have a shop to build this new MiniViggen, which actually turned out to be the VariEze. And my plan was, hey, I can't make much money probably on the VariViggen. It's too hard to build. I won't have a lot of people, but if I make something that's real easy to build, I can sell hundreds of sets of plans."

As it turned out, he would sell many more sets of plans than that.

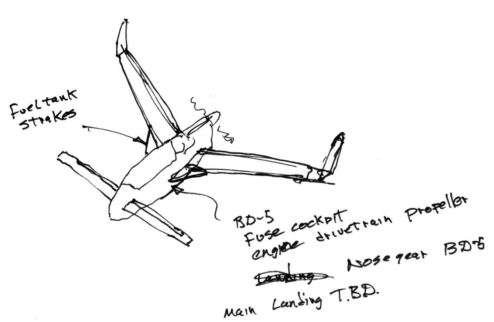

Not a lot of information remains about the MiniViggen, which is shown in this recent drawing sketched by Burt Rutan in December 2009. The MiniViggen was originally sketched in 1971, before Rutan joined Bede Aircraft. One early drawing even shows the wingtips pointed downward. The VariViggen was already flying, but he intended the MiniViggen to be his first homebuilt design. *Courtesy of Burt Rutan*

Building with Composites

A composite is simply a combination of two or more materials. The reason for such a combination is to take advantage of the benefits that each individual material has to offer.

Like the composite plywood where layers of wood are glued together at different orientations, sheets of fiberglass can be bonded together by epoxy to form a very stiff composite.

Fiberglass is strong but dense. Polystyrene, commonly known as Styrofoam or just foam, is lightweight but not strong. Making a full-sized aircraft entirely out of one of these materials or the other wouldn't be a very good idea. But by wrapping fiberglass around foam, the resulting structure is both strong and lightweight.

"I wandered over to the guy who was repairing the European fiberglass sailplanes and watched him," Rutan said, having had his first taste of composites with the elevator of the canard he built for the MiniViggen.

Rutan sought to learn more about this method of construction from Fred Jiran Glider Repair. He already had an idea that he could make a wing with composites without the need for all kinds of specialized tooling, which were simply molds as used in composite work.

"I watched him repair these without using the tooling," Rutan said. "I thought wait a minute, maybe I can build an airplane the way he's repairing sailplanes and never even have a tool. I envisioned the homebuilders would buy blocks of foam and rolls of fiberglass from Aircraft Spruce, and nobody would ship a tool. Let's see if I can build this? I found to my delight that not only was it easy to build, but I reasoned, correctly, that it was a lot more reliable for somebody to do this with inspection criteria that I gave than it was for him to take 4130 steel and weld it to make a fuselage. You make a bad weld, you die. Here, a bad part, you could see it with fiberglass. It's robust. It's extra strong."

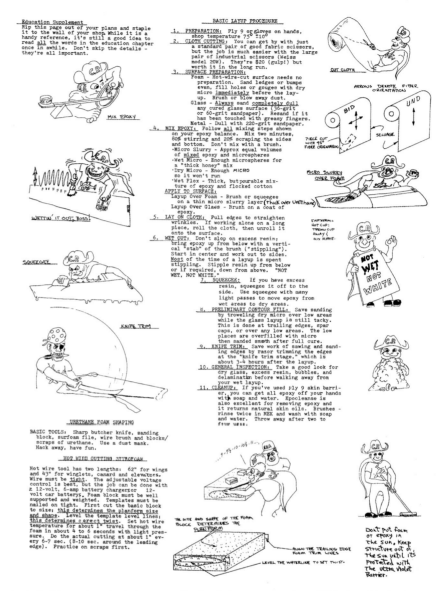

Included in each set of plans was the basic layup procedure. With the aid of cartoons drawn by Gary Morris, Burt and Carolyn Rutan's first employee of RAF, this step-by-step process described the materials to use and how to go about building the composite structure needed for Rutan's homebuilt airplanes. *Courtesy of Burt Rutan*

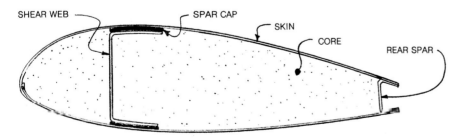

The cross-sectional view of a wing built using composites is shown. Foam forms the core that fills most of the inner volume, keeping the wing rigid and lightweight. To give the wing strength, the shear web, spar cap, skin, and rear spar are made from fiberglass plies bonded together with epoxy. *Courtesy of Burt Rutan*

When Burt Rutan gave talks about his aircraft or workshops about building with composites during the Experimental Aviation Association's annual air show in Oshkosh, Wisconsin, it wasn't difficult to tell which tent he was speaking from. *Courtesy of Burt Rutan*

A New Method of Construction

Rutan had already hotwired wing cores out of foam for some of his model airplanes. The hotwire process simply uses an electrically heated wire to cut through foam, like a hot knife through butter.

"I thought, well, I'll try to hotwire on two-pound foam instead of just one-pound foam," Rutan said, where one-pound and two-pound were just measures of density. "Two-pound foam was this blue stuff that's used for flotation on docks and piers and so on. The one-pound is made for picnic coolers and packaging and whatever. The one-pound worked fine for model airplanes, but I reasoned that the two-pound wouldn't be too heavy to do a full core wing for a VariEze."

Known as the layup process, the shaped foam is first coated with room-temperature curing epoxy and then covered with a sheet of fiberglass cloth. As the strength of a wood board varies depending on the orientation to its grain, the strength of fiberglass cloth depends on the orientation to its weave. After the initial fiberglass cloth is applied, a bonding layer of epoxy is again added and another sheet of fiberglass cloth is placed at a specified angle that crisscrosses the orientation of the initial fiberglass cloth. A number

of subsequent layers of epoxy and fiberglass cloth are applied in this fashion as required by the structure's design.

The type of foam also depended on application: polystyrene foam in wing cores, PVC foam in fuselage bulkheads, and urethane foam in the fuselage and fuel tanks.

Although the use of composites would allow Rutan to build safe, strong, affordable, and lightweight structures, these weren't the most attractive features to him. Composites would allow him to build complex shapes very simply and quickly. This, in turn, would reduce the number of required parts. Rutan could then get aircraft built quickly and into flight testing sooner. Whereas the wood and aluminum VariViggen took four and half years to complete, his next aircraft would take a mere three and half months.

The types of composites and the methods of fabrication did vary by some degree between Rutan's designs. Though he started with fiberglass and foam, honeycomb structures and composites made with carbon fiber were used for requirements of very high strength and very low weight, as needed by the record-breakers Voyager and SpaceShipOne.

Winglets

Inspired by the flight of soaring birds, NASA researcher Richard Whitcomb invented the Whitcomb winglet as a way to improve the aerodynamic efficiency of a wing. Whitcomb termed them winglets because he wanted to highlight their winglike, airfoil shape.

In normal flight, induced drag is generated as high pressure air from below the wing mixes with low pressure air from above the wing. A major source of an aircraft's induced drag is air flow around the wingtip from underneath the wing to above it, resulting in a wingtip vortex. By using a Whitcomb winglet to mitigate this effect, Whitcomb found a 20 percent reduction in induced drag and a 9 percent increase in the lift-to-drag ratio. This translates directly into fuel savings.

Rutan learned of Whitcomb's new discovery in 1974 as he designed the VariEze proof of concept (POC).

"It doesn't put a lot of extra bending moment on the wing, but it helps the induced drag," Rutan said. "It's got to be shaped just right, and there's a fin on the top and a little fin on the bottom. Whitcomb had tested it in a wind tunnel. He was hoping that someday somebody would put it on a Learjet or an airliner. He swore that it would give more range."

The design for VariEze POC had already called for vertical fins on the wingtips. By making them Whitcomb winglets, Rutan could add some extra aerodynamic efficiency to the aircraft.

"Looking at drawings and reading what he said about the lift coefficients that he gets top and bottom,

they are set up so they tend to unwind the normal rotation of the wingtip vortex," Rutan said. "I figured that out. And I made them part of the original VariEze POC design. From the time I found out about them until the first flight of the VariEze POC was only a few months. So I was the first to fly winglets."

However, when Whitcomb got wind of Rutan's application of the winglets on the VariEze POC, he was critical and dismissed their use for light aircraft, figuring Rutan didn't design them correctly.

The following year, Whitcomb attended Oshkosh.

"I dragged him out to the flightline. Now, I'm showing him the homebuilt VariEze, which has a lot better version of the Whitcomb winglet," Rutan said, comparing the homebuilt VariEze (Model 33) to the smaller VariEze POC (Model 31).

Whitcomb inspected the design, and, to his delight, he found Rutan had built the winglets properly. Whitcomb told Rutan he was proud to see them on the aircraft and that Rutan was getting an additional 5 percent range with them as opposed to without them.

"I finally got his favor," Rutan said. "That was a big thing for me in 1976."

Although significantly researched by general, commercial, and military aviation in the late 1970s, implementation had been slow, and it had taken several decades for the use of the efficiency-improving Whitcomb winglets to be widely adopted by the industry.

Burt Rutan's VariEze POC was the first airplane to ever fly with the Whitcomb winglet, as shown here. These winglets were not fins, but both the top and bottom parts were actually airfoils, in the shape of a wing. The winglets counteracted the drag normally created by wingtip vortices. Rutan had the winglets even double as rudders. *Courtesy of Burt Rutan*

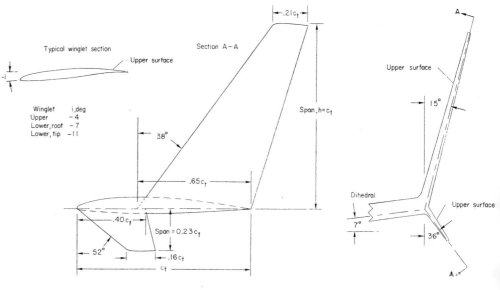

In 1974, Burt Rutan came across new research by leading NASA aerodynamicist Richard Whitcomb. The research showed that by using a winglet attached to the tip of the wing of an airplane, it would reduce drag and improve range. Only a few months away from the first flight of the VariEze POC, Rutan decided to incorporate the winglets in his new all-composite aircraft. *NASA*

VariEze POC

The construction of both the completed VariViggen and the unfinished MiniViggen presented problems for Burt Rutan. These aircraft were difficult to build. Now, however, he had figured out a method using composites that would open up huge possibilities. Designs with shapes too complex for practical wood and metal fabrication could now be considered.

Rutan was convinced that by using these new composite construction methods, with existing materials, he could dramatically simplify the building process, provide high performance and high efficiency, and create a jaw-dropping, futuristic look. A homebuilder's dream. So he set his sights on an aircraft that could smash existing speed and distance records for a piston-engine aircraft weighing between three hundred and five hundred kilograms.

The design for the VariEze proof of concept (POC) began shortly after Rutan relocated to Mojave in 1974. A POC is a prototype that is a testbed and not likely to be mass produced. Rutan had intended to use the VariEze POC to research and stretch the boundaries of construction, performance, and efficiency. It had to be *very easy* to build, and that's how the name originated. Construction was completed in three and a half months with the help of Gary Morris, who came by in the evenings after working at Fred Jiran Glider Repair. Rutan would take him to dinner afterwards. Morris later quit his day job and got paid to work at RAF.

Features of the VariEze POC included the fiberglass and foam composite structure; a canard with elevons to control both the pitch and roll; high aspect-ratio, sweptback wings; Whitcomb winglets on the main wing

The VariViggen (bottom) was mostly wood with aluminum wings, and the MiniViggen was planned to be all metal. Building with composites allowed Rutan to build complex curved shapes much easier and faster than working with wood or metal. It took Rutan four and half years to build the VariViggen, but it took only three and half months to build the VariEze POC (top). *Courtesy of Burt Rutan*

First flown in 1975, the VariEze proof of concept (POC), Model 31, was an enormous jump in design compared to the VariViggen. The use of composites allowed Burt Rutan to build an incredibly high performing and efficient aircraft. It was named VariEze because it was *very easy* to build. The name VariEze was suggested by Rutan's sister, Nell. *Courtesy of Burt Rutan*

that also contained the rudders for yaw control; tandem seating; a pusher engine; and a wide CG range due to the canard configuration.

The VariEze POC began flying in May 1975. With its 62-horsepower Volkswagen automotive engine, the VariEze POC reached a top speed of more than 180 miles per hour. At cruise speed, it achieved a fuel efficiency of more than 40 miles per gallon, and at economy speed, it exceeded 60 miles per gallon. These numbers encouraged Rutan to kit the aircraft. He determined that a kit of the VariEze POC would take 350 hours to build.

First World Record

By July 1975, the aircraft was ready to make its way to EAA's fly-in and convention. Burt Rutan and his brother Dick Rutan, a top USAF fighter pilot, planned to go for a world record for closed course distance at Oshkosh. The current Class C-1a world record holder, Ed Lesher, covered 1,554 miles flying his Teal Airplane in 1970.

Dick Rutan set out to fly to Oshkosh from Mojave nonstop in the VariEze POC with an extra fuel tank in the backseat while Burt Rutan flew out earlier in the VariViggen, which had to stop for fuel along the way. But hot engine oil temperature and low engine oil pressure forced the VariEze POC to land before reaching Oshkosh.

After repairs and a few days of wowing the crowds during the air show, Dick Rutan took off in the VariEze POC on a Saturday to break the world record. While only on the second lap of the 182-mile circuit, the engine blew and Dick Rutan had to make a deadstick landing in Green Bay, Wisconsin.

"We recovered the airplane, trucked it back to Oshkosh, and looked around the field," Burt Rutan said. "Here is John Monnett, who was using a Volkswagen engine. It looked like a better one. And I asked him, 'Do you have any engines around?'

"He said, 'No, I have them down in Chicago where I live.'

"And I said, 'I want to buy one from you.' And we went down there in a station wagon, bought this engine, and brought it back up. And on the flightline at Oshkosh, out in the open, we swapped out and put this John Monnett engine in it."

By Sunday, the engine had been replaced and flight tested. On Monday morning, Dick Rutan lifted back into the sky to push the endurance of the VariEze POC. Nine laps and 13 hours, 8 minutes, and 45 seconds later, he touched down and set a new world record at 1,638 miles. During the flight, the VariEze POC averaged a fuel efficiency of 3.1 gallons per hour or 40.7 miles per gallon while flying at an average speed of 125.5 miles per hour.

Because of the thin wing on the VariEze POC, there wasn't much wing area to carry all its weight. But this high wing loading made for a fast moving airplane, capable of flying more than 180 miles per hour. Cooling for the VW engine came from a streamline air scoop on the belly between and slightly behind the rear landing gear. *Courtesy of Burt Rutan*

The VariEze POC had fixed rear landing gear and manually retractable nose gear. This lowered the drag and kept the weight down. The airplane parked on its nose, kneeling on a reinforced section of the fuselage. This helped entry into the cockpit and reduced the need for tie down. *Courtesy of Burt Rutan*

Dick Rutan, shown seated in front with his brother Burt Rutan in back, piloted RAF aircraft on many important flights. Only about two months after the VariEze POC made its first flight, he soloed it cross-country to Oshkosh, Wisconsin, for the EAA's air show. While there, he broke a long-distance world record for flying 1,638 miles in a little over thirteen hours. *Courtesy of Burt Rutan*

The Need for Improvements

Ultimately, Burt Rutan was disappointed by the VariEze POC's performance. He had intended the aircraft to have a much better range. There were other issues as well. Early airflow separation on the canard resulted in a high stall speed of 60 knots. And below 80 knots, the aircraft exhibited poor roll control. It was time to return to Mojave.

"We had only one hundred hours on the airplane, and we had failed an engine during this record attempt," Rutan said. "Now flying it home with a Monnett engine, when we got to Phoenix that engine failed."

No doubt the VariEze POC provided plenty of opportunities for learning and understanding.

"I am planning on kitting this airplane," Rutan had told everyone while at Oshkosh. "It's going to be real low cost. It's got a Volkswagen engine. You won't have to buy an aircraft engine. Look at it. It goes 170 miles an hour, burns only three gallons per hour. And it's two-place."

Rutan had already proven that he could sell plans and put a kit together for homebuilders, and in many ways

The three "EZ" prototypes, the VariEze POC (left), Long-EZ (center), and VariEze homebuilt (right), appear to be quite a similar. However, there were significant differences between them in size, handling, speed, and range. The VariEze POC was the only one of the three never offered as a homebuilt. *Courtesy of Burt Rutan*

the VariEze POC had proven to be an advancement and a significant leap up from the VariViggen.

"I got home after the summer of '75, but now I had two engine failures. I've got an airplane with pretty high wing loading for real high speed. And you know, it isn't as easy to fly as a Cessna or some of these little biplanes or low-wing loading or low-stall speed things."

With the help of Dick Eldridge at NASA's Dryden Flight Research Center, Rutan found that a canard using a GU25 airfoil should replace the VariEze POC's existing canard, which had a GAW-1 airfoil. After the replacement, the stall speed improved from 60 knots to 52 knots. This switch also rectified the roll control problem. However, Rutan could not find a suitable engine to fit inside the VariEze POC that could replace the unreliable Volkswagen engine.

"I got scared about people having engine failures," Rutan said. "So I blew the whistle on my whole plans and said, 'I'm going to build an airplane that will use an aircraft engine. It will be bigger. It will have some room for baggage at least. It will have more range. It will be reliable. It will just be the right thing ethically to put out to encourage people to build it and fly around with their families.' "

VariEze POC Details

Model number	31
Type	single-engine, canard pusher
Prototype tail number	N7EZ
Current prototype location	EAA AirVenture Museum, Oshkosh, WI
Customer	RAF R&D
Fabrication	RAF
Flight testing	RAF
First flight date	21 May 1975
First flight pilot	Burt Rutan
Seating	two-place, tandem
Wingspan	21 ft
Wing area	59 ft² (canard and main wing)
Length	12.4 ft
Empty weight	399 lbs
Gross weight	880 lbs
Engine	Volkswagen 1,834 cc, 62 hp
Landing gear	tricycle, retractable nose gear & fixed main gear
Fuel capacity	14 gal
Takeoff distance	700 ft
Landing distance	1,100 ft
Rate of climb	1,100 fpm (gross weight)
Max cruise speed	173 mph (gross weight)
Range	780 miles (gross weight, 40% power)
Ceiling	14,000 ft (gross weight)

With a new method of construction available to Burt Rutan, he revisited his first airplane, the VariViggen (Model 27). He eliminated all aluminum construction from the aircraft by redesigning the wing and rudder out of foam and fiberglass. This new design, Model 32-SP, was called the VariViggen Special Performance (SP). *Courtesy of Burt Rutan*

VariViggen SP

"Here I was building—I don't know how I had this much time—I was building the VariEze POC. Hadn't flown yet," Burt Rutan said. "And I thought, 'I built the wings on the VariViggen out of metal just because I wanted to learn how to build with metal.' It was a wooden airplane with metal wings. It made no sense. But that's the way it always flew. I figured out that just by stretching the wings a few feet on each side, but making the sweep different, I could make it balance and have the same controllability. It would have a higher aspect ratio. It would perform better at altitude, have longer-range, and in these outer wings I would put fuel tanks."

The VariViggen had a single, gravity-fed fuel tank behind the passenger seat and above the baggage compartment. The tank's capacity was thirty gallons, of which five gallons were reserve, giving the aircraft a range of three hundred miles. This had been too short for Rutan. To get around this, while still at Bede, he took a propeller spinner without the propeller cutouts and attached it to a sheetmetal cylinder fashioned out of aluminum.

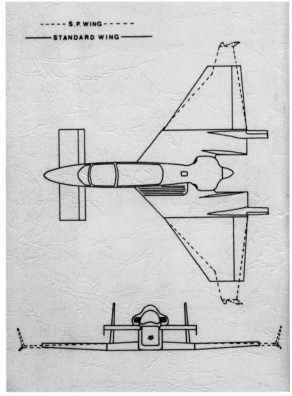

These planform and head-on views show how the aluminum wing of the VariViggen was redesigned for the VariViggen SP. With the use of composites, Burt Rutan also changed the flat bottom wing to a more aerodynamically sound contoured shape as well as added a Whitcomb winglet. *Courtesy of Burt Rutan*

By also taking advantage of lessons learned from building and flying the VariEze POC, Burt Rutan improved some of the VariViggen's features by designing a retrofittable special performance wing. The gross weight of the VariViggen SP did not change because the use of lightweight composites allowed it to carry an additional 18 gallons of fuel. *Courtesy of Burt Rutan*

"We were building BD-5s, so we knew how to do aluminum," Rutan said. "I made something that looked like a bomb. It held twelve gallons of gas, and I had a little pump in it that pumped up into the other tank. Made a world of difference because you have a certain amount of unusable fuel, in other words, reserve fuel. If you add twelve gallons, you still have the reserve fuel. So I added like 50 percent to the range by just putting this twelve-gallon tank on. And it looked cool. It was my fighter. It had a centerline bomb. I flew almost all the time with that bomb on it."

Remember the good old days when you could fly around with something that resembled a bomb strapped to the belly of your airplane? Decades later Mike Melvill encountered trouble in White Knight while delivering SpaceShipOne to the Smithsonian's National Air & Space Museum. SpaceShipOne had looked too much like ordnance as well, which caused a stir on the ground as he tried to land in Washington, D.C.

Although the centerline fuel tank extended the range to 450 miles, Rutan never made a homebuilders version of it.

A New Pair of Wings

Fabricated using fiberglass and foam and then first flown on 16 July 1975, a month after the first flight of the VariEze POC, the VariViggen special performance (SP) wings simply swapped places with the detachable, outer metal wings. With the SP wings in place and no other modifications, the VariViggen was then designated as the Model 32-SP. The gross weight didn't change, even though the composite construction was much lighter. The SP wings took advantage of Whitcomb winglets, which were first used on the VariEze POC, and had built-in fuel tanks that each contained nine gallons.

"It wasn't as much fun to fly as it had smaller ailerons," Rutan said. "It didn't have that real zippy roll rate. But it had a little flatter approach with longer wings. It had a better glide. And it had maybe 5 or 8 percent more range. If you go up real high, it would perform better, maybe get another 10 percent up real high. Of course, you didn't go very high in a VariViggen. It pretty much wouldn't climb well if you get above all eleven thousand feet or so."

The stall characteristics with the SP wings also changed. Not for the better. With the standard wings, the VariViggen was "stall proof," but with the SP wings, it behaved more like a conventional light aircraft. This fact may have been a step backward, but, overall, the VariViggen matured to a more practical aircraft. True it didn't roll as fast, but it did climb and cruise faster.

With the boost from the more efficient wing design and the extra eighteen gallons of fuel, the maximum range of the VariViggen now increased to more than six hundred miles.

Rutan sold plans for the SP wings before selling plans for the VariEze homebuilt. Unlike when Rutan built the VariViggen, this time he knew from the start that he wanted to sell plans for the SP wings. He did a much better job of photographing and documenting the construction. The plans now included templates that homebuilders used to hotwire carve the foam wing cores. Two crossed layers of unidirectional fiberglass cloth were required for the skin. Compared to the metal wings, the SP wings took a third of the time to build.

VariViggen Wing Comparison

	Standard Wing (Model 27)	SP Wing (Model 32-SP)
Wingspan	19 ft	23.7 ft
Wing area	119 ft²	125 ft²
Aspect ratio	3.03	4.47
Gross weight	1,700 lbs	1,700 lbs
Rate of climb	800 fpm	1,000 fpm
Cruise speed	150 mph	158 mph
Range	300 miles	600 miles

The VariEze homebuilt (Model 33) was the first aircraft to hit the homebuilt market from Burt Rutan where his original intention—from start to finish—was to design an aircraft specifically for homebuilders that would be very easy to build and have great performance. A major difference between the larger homebuilt as compared to the POC was the use of a certified aircraft engine that replaced the modified Volkswagen engine, which was prone to failure and could lead to accidents. *Courtesy of Burt Rutan*

While maneuvering, aircraft can easily experience g-forces greater than 1 g. This static load test for the homebuilt VariEze uses sandbags to simulate the maximum g-force the canard can safely handle. *Courtesy of Burt Rutan*

VariEze Homebuilt

Up to this point, Burt Rutan had designed, built, and flown the VariViggen (Model 27 and Model 32-SP) and the VariEze POC (Model 31). As of January 1975, an estimated 150 VariViggens were under construction. Many of them would remain unfinished because of the aircraft's complexity, though. His new composite construction method seemed to solve that problem with the VariEze POC. However, from the experience Rutan gained from these two aircraft, he felt the performance of the VariEze POC was inadequate to develop into an aircraft for homebuilders. So Rutan focused his efforts on the bigger and better VariEze homebuilt.

At first glance, the VariEze POC and the VariEze homebuilt look to be the same aircraft, but that is about where the similarities end. Rutan outlined his design philosophy for his new airplane in an article he wrote for *Sport Aviation* in January 1976. He stated:

Every aircraft is designed by a series of compromises to fit a list of requirements. In the design of the homebuilt VariEze, I specified two place plus baggage and a cruising range of over 800 miles. I then listed the order of priority of the remaining requirements as follows:

1. *Efficient cruise at a relatively high speed.*
2. *Simplicity of construction obtained by few number of parts.*
3. *Low maintenance requirements of systems and structure.*
4. *High structural life and system reliability.*
5. *Flying qualities optimized for low fatigue on cross country flights.*
6. *Cockpit comfort.*
7. *Good riding qualities in turbulence.*
8. *Low cost.*
9. *Light weight for good climb performance.*
10. *Ease of disassembly for trailering.*
11. *Low stall speed.*
12. *Short takeoff/landing distance.*
13. *Aerobatic capability.*
14. *Soft/rough field capability.*

This is not to say that the last few items were ignored completely, but they were not optimized at the expense of the first few items.

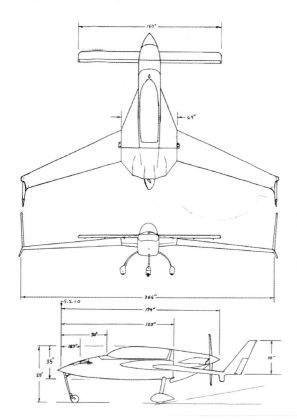

During the flight testing phase, to improve yaw control, an experimental rhino rudder was attached to the nose of the VariEze homebuilt. While effective, it did sit directly in front of the pilot's forward view. The idea was nixed, and the rudders stayed put in the winglets.
Courtesy of Burt Rutan

Using the VariEze POC as a baseline, longer range was a major design requirement for the VariEze homebuilt. It had a range of 1,050 miles, which was 270 miles greater than that of the VariEze POC.
Courtesy of Burt Rutan

After scrutinizing this ambitious list of design criteria, Rutan developed the revolutionary VariEze homebuilt. With its performance, flying qualities, stall resistance, affordability, and ease of construction, nothing at the time could come close to matching it in the air. Even today, as of 2010, VariEze homebuilts hold several distance and speed world records.

The sleek surface finish of the VariEze, which produced very little air resistance, and the highly aerodynamic shape made for one hot aircraft. With its 100–horsepower engine, carrying limited fuel and no passenger, it hit a speed of 200 miles per hour. An altitude of 25,300 feet had also been reached. Klaus Savier currently holds a world speed record of 203.4 miles per hour in his VariEze.

Unlike the VariEze POC, which had elevons on its canard to control both roll and pitch, the VariEze homebuilt required separate ailerons on the main wing for proper roll control. Elevators on its canard were used for pitch. The ailerons and elevators were controlled with a joystick on the right side of the cockpit. Rudders were mounted in the Whitcomb winglets, and only the front-seater had rudder pedals to control them.

All Dressed Up

If an aircraft is said to be easy to build, well, that is really a relative thing. Not many people would actually ever attempt such a feat in the first place. An aircraft designer with all the information in his or her head about a shiny new prototype must be able to communicate the assembly process to the would-be homebuilders so there isn't a whole bunch of leftover nuts and bolts next to a squirrelly constructed airplane. Most everyone has experienced the frustration of trying to put something together with a lousy set of instructions.

Rutan realized that his plans for the VariViggen were not ideal. Kitting the VariViggen was, after all, an afterthought. By designing the VariViggen SP wings, he also got to practice making much more effective plans and took this learning straight into drawing up the plans for the VariEze homebuilt.

"It was a different deal once I learned more about the business," Rutan said. "My new plans were based on the Simplicity dress patterns used to make dresses. If you get a Simplicity dress pattern, there are a few words and a sketch and then a few more words and another sketch. Each sketch shows you graphically what that paragraph tells you to do."

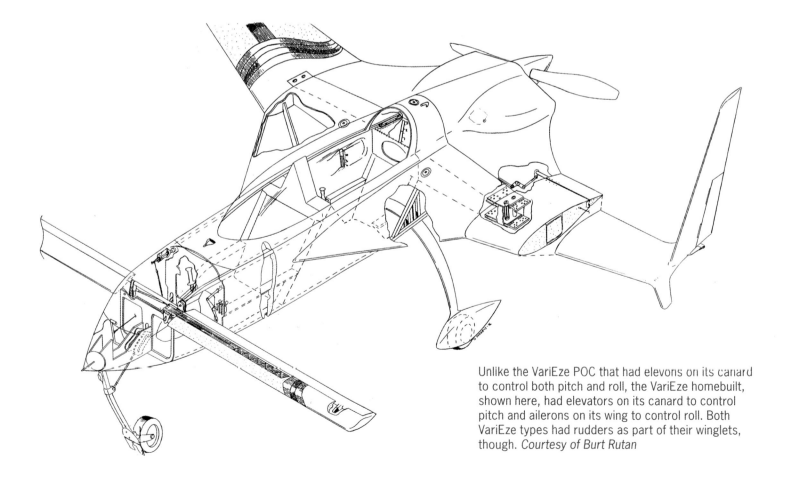

Unlike the VariEze POC that had elevons on its canard to control both pitch and roll, the VariEze homebuilt, shown here, had elevators on its canard to control pitch and ailerons on its wing to control roll. Both VariEze types had rudders as part of their winglets, though. *Courtesy of Burt Rutan*

Never one to shy away from new technology, in the April 1977 issue of *Canard Pusher*, Burt Rutan included a short description and a sketch of a solar-powered electrical system for the VariEze homebuilt that he would soon be testing. A rectangular bank of solar cells can be seen directly in front of Rutan in the photo. *Courtesy of Burt Rutan*

Below: Built around a bigger and more powerful Continental aircraft engine as opposed to a Volkswagen automotive engine, the VariEze homebuilt (bottom) was longer by 1.8 feet, had a wingspan wider by 1.2 feet, and a gross weight 170 pounds heavier than the VariEze POC (top). *Courtesy of Burt Rutan*

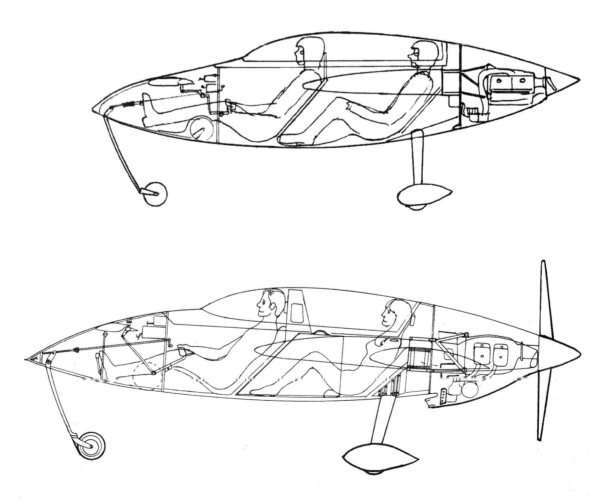

All Burt Rutan supplied were plans to make his VariEze homebuilt. Individuals who purchased the plans built their own airplanes and then flew them. After Rutan introduced the plans in Oshkosh 1976, in 1977 there were five VariEzes at Oshkosh. In 1978, there were twenty-four and in 1979 forty-one. EAA even assigned a special area to park for those who made the annual trip to Oshkosh in their VariEzes and Long-EZs, which would soon follow. *Courtesy of Burt Rutan*

The common practice during those days was to provide a big roll of blueprints and a set of instructions. However, Rutan's plans now consisted of five sections: a 1-inch-thick, 11- by 17-inch manufacturing manual, engine installation manual, finishing instructions, owner's manual, and optional electrical instructions. The RAF's quarterly newsletter *Canard Pusher*, originally called *VariViggen News*, updated the homebuilders on plans changes and gave helpful building hints.

"You show them only what they need to know for what they're doing today or tonight," Rutan said.

So the format of the plans for making his VariEze and subsequent homebuilt airplanes was based on the plans for making dresses. It didn't take a homebuilder to be an expert craftsman, and someone with a good set of hands could build a VariEze in six hundred to one thousand hours. This style of plans became widely emulated by other homebuilt aircraft designers.

On the Road

Nowadays the use of composites is widespread. Look no further than modern houses, bicycles, tennis rackets, skies, or fishing poles. But in the 1970s, composites were still a mystery material to many. Homebuilders initially approached Rutan's composite-building method with some trepidation. Rightfully so. Flying well over one hundred miles per hour at five thousand feet is a terrible place for homebuilders to learn that they jumped into something that they didn't fully understand.

Rutan decided it was up to him to teach people how to build with composites. Once people understood composites and felt comfortable building with composites, people could then get to work building his futuristic little airplane.

"I took trips to France, England, Netherlands, Germany, Australia, New Zealand, and all over the United States," Rutan said. "I brought along materials to show people how to hotwire, how to do fiberglass layups, how to do knife trims, how to do sanding for bonding, and all the processes."

He ran forums at Oshkosh and weekly Saturday demonstrations from his shop in Mojave.

"I thought I had to show people. I thought if I could teach thousands of people to do this, then there would be enough people around that they could help to teach others locally. And this process of building airplanes, which was brand new, would then spread educationally. It would support the plans sales that looked like were going to happen. This was the big thing and was so important to the success of the VariEze."

In the Hands of Homebuilders

In the late 1970s, to assemble a VariEze would cost between five thousand and nine thousand dollars. Homebuilders had the option of building mostly from scratch or buying a bunch of prefabricated parts. They also could install a new or used engine. Not only was the VariEze cheap to build, it was cheap to operate and maintain.

The plans went on sale in July 1976. A year later, five VariEzes made the pilgrimage to Oshkosh. In 1978, there were twenty-four, and in 1979 there were forty-one.

Although not unexpected because of the different skill level each homebuilder had, there was a disparity in performance between their aircraft. For such an aerodynamically clean aircraft, small variations in the surfaces that were exposed to the airstream could have large effects on performance. One survey conducted by Rutan revealed the average VariEze performance was twelve miles per hour less than expected. Many homebuilders couldn't fight the urge to pack in extra equipment and instrumentation as well. Designed to be a lightweight aircraft, these additions caused weight gain that substantially decreased the useable payload.

Homebuilders did, however, play an important role in shaking out the design. As they encountered problems, Rutan quickly began investigating, made the necessary modifications, and sent out alerts in *Canard Pusher*. By 1980, homebuilders purchased more than three thousand sets of plans at $128 a pop, and a squadron of more than two hundred VariEzes had taken to the skies all around the world.

VariEze Homebuilt Details

Model number	33
Type	single-engine, canard pusher
Prototype tail number	N4EZ
Current prototype location	National Air & Space Museum, Washington, D.C.
Customer	homebuilders, marketed 1976
Fabrication	RAF
Flight testing	RAF
First flight date	14 March 1976
First flight pilot	Burt Rutan
Seating	two-place, tandem
Wingspan	22.2 ft
Wing area	66.6 ft² (canard and main wing)
Length	14.2 ft
Height	4.9 ft
Empty weight	585 lbs
Gross weight	1,050 lbs
Engine	Continental O-200, 100 hp (4 cylinders)
Landing gear	tricycle, retractable nose gear & fixed main gear
Fuel capacity	28 gal
Takeoff distance	860 ft (gross weight)
Landing distance	1,000 ft (gross weight)
Rate of climb	1,500 fpm (gross weight)
Max cruise speed	193 mph (gross weight)
Range	1,050 miles (gross weight, 40% power)
Ceiling	20,500 ft (gross weight)

This photograph from Oshkosh in 1982 shows nine original Rutan types and two derivatives (Gemini and Cozy). All from left to right: in the front row, Solitaire, Grizzly, and Gemini; in the middle row, AMSOIL Biplane Racer, AD-1, VariEze homebuilt, and Quickie; and in the back row, Cozy, VariViggen, Defiant, and Long-EZ. The Gemini was a two-place twin based on the four-place twin Defiant, and the Cozy was a side-by-side seater based on the tandem seater Long EZ. *Courtesy of Burt Rutan*

Chapter 2

Rutan Aircraft Factory Continues to Soar

In June 1974, Burt Rutan had settled back in California, deciding to go solo and set up shop in a wooden World War II barracks at Mojave Airport. And within two years, he had the VariEze POC and VariEze homebuilt flying.

Rutan had founded the Rutan Aircraft Factory, which was originally called the Rutan Aircraft Company, back in 1969 while he worked at Edwards Air Force Base and built his first airplane, the VariViggen, in his spare time.

RAF sold a couple of VariViggen plans a week, but after the VariEze homebuilt flew to Oshkosh in July 1976 and its plans hit the market, business lifted way off.

"People had been waiting to buy them for months, and I did not accept advance orders. The first day I sold one hundred sets. A guy flew in from San Diego and bought about twenty sets for him plus his friends. And they cost $139. Well, one hundred times one hundred is ten thousand dollars—in a day. Wow, pretty darn good return." Rutan said. "That year I took home six thousand bucks for the family. We didn't buy a new car or anything. You could live on six thousand bucks in Mojave in '76. I put the rest in the bank, so we could build another airplane and grow the company."

When Rutan looked back, he felt that his flight test experience for the air force was crucial to learning how to efficiently analyze and take risks as he developed his own designs. His subsequent exposure working for Bede gave him insight into the business side of aviation. Now, on average, every year one new type of airplane rolled out of Rutan's hangar and took off.

"A lot of these airplanes were done just because I was pissed off at some critique," Rutan half admitted with a smile. "And I thought, 'Well, let's see if this guy is right.'" This was certainly the case when someone said that he couldn't get a low landing speed with a canard—then came Grizzly.

That was exactly the spirit Rutan charged through aviation with. And as he did so, Rutan's design and fabrication work received notice from more than homebuilders.

Aside from doing his own new aircraft, such as the Defiant and Long-EZ, he was being asked to help design and fabricate aircraft for others, such as Quickie and AD-1. Some projects took longer to go from first concept to first flight. The AD-1, for example, was Model 35, but it made its maiden flight after those of the Defiant (Model 40), Quickie (Model 54), and Long-EZ (Model 61).

In 1977, RAF expanded into a new building and growth continued.

"Mike Melvill came on board in 1978. He helped me finish the Defiant. I was building the sump tanks for the Defiant. I wasn't finished with this twin. He helped me finish it. He had already built a VariViggen. His big job was to do builder support, but he couldn't do builder support on the VariEze because he never built one. So I immediately had him build his own airplane."

RAF sold approximately fourteen thousand sets of plans. With all those builders out there who had either finished, started, or not yet begun building their airplanes, it was necessary for Rutan to disseminate important information to them, such as plan updates, building techniques, and safety warnings. Started in May 1974 as *VariViggen News*, RAF's quarterly newsletter was renamed after five issues to *Canard Pusher* to better reflect its coverage of a whole fleet of designs.

Quickie

"When the development of a new homebuilt aircraft is undertaken, it is often closely followed by the public, and can even be viewed on static display at fly-ins and air shows during its construction period, at which time its performance and cost estimates are echoed by the developers," reported Burt Rutan, Tom Jewett, and Gene Sheehan in *Sport Aviation*. "This is not the policy of our little skunk works at the Mojave Airport. The development of the Quickie has been one of the best kept secrets of aviation. Until its first flight on November 15, 1977, its existence was known only to a handful of people."

In 1975, Jewett and Sheehan started searching for a small and reliable engine that they wanted to build a single-place kit aircraft around. They wanted the aircraft to be very enjoyable and economical. They chose an Onan engine that was actually used to provide electricity for mobile homes.

Jewett and Sheehan turned to Rutan for the preliminary design in May 1977. The first concept, Model 49, looked like a miniature VariEze, but the configuration didn't work aerodynamically. The final design, Model 54, solved those problems. Head-on, the Quickie looked like the X-wing starfighter from *Star Wars*, which just so happened to have its silver screen release that same month.

From the side, the aircraft looked like a biplane with the bottom wing pushed forward and the top wing pushed backward. This configuration is more commonly know as a tandem wing or stagger wing. To improve stability, the top wing canted up and the bottom wing canted down.

The full-span elevators were mounted on the bottom wing, and the front landing gear was housed in its wingtips. Inboard ailerons on the top wing controlled the roll. The tail fin was fixed for directional stability, but the small rudder, mounted to the tail wheel, provided yaw control.

Using the same composite building method, the Quickie weighed half as much as the VariEze, an incredible 240 pounds empty, and it took two-thirds of the time, materials, and cost to construct. At a maximum cruise speed of 121 miles per hour, it had a fuel efficiency of 80 miles per gallon, but it hit 100 miles per gallon when the speed dropped down to its efficiency cruise speed.

After Rutan, Jewett, and Sheehan had finalized the design and the Quickie was built, Rutan completed the flight testing and then turned the aircraft over to Jewett and Sheehan's Quickie Aircraft Corporation to begin selling kits. The Quickie was the first aircraft to take flight that was contracted by a customer for design by Rutan.

Tom Jewett, a former coworker of Burt Rutan's at Bede Aircraft, and Gene Sheehan approached Rutan to help design a superefficient, zippy, single-place aircraft. Phenomenally lightweight, the Quickie had an empty weight of 240 pounds but could fit a six-and-a-half-foot-tall, 215-pound person inside. *Courtesy of Burt Rutan*

Quickie Details

Model number	54
Type	single-engine, tandem-wing
Prototype tail number	N77Q
Current prototype location	The Museum of Flight, Seattle, WA
Customer	Quickie Aircraft Corporation
Fabrication	RAF
Flight testing	RAF
First flight date	17 November 1977
First flight pilot	Burt Rutan
Seating	single-place
Wingspan	16.7 ft
Wing area	53 ft²
Length	17.3 ft
Height	4 ft
Empty weight	240 lbs
Gross weight	480 lbs
Engine	Onan, 22 hp
Landing gear	conventional, fixed
Fuel capacity	8 gal
Takeoff distance	660 ft
Landing distance	835 ft
Rate of climb	425 fpm
Maximum speed	127 mph
Cruise speed	121 mph
Range	550 miles
Ceiling	12,300 ft

The Quickie had full-span elevators on its bottom wing and inboard ailerons on its top wing. Originally, as shown here, the vertical stabilizer was fixed and the tail wheel pant, at the way back, contained the rudder. A subsequent modification enlarged the vertical stabilizer and included the rudder in it. *Courtesy of Burt Rutan*

The tandem-wing Quickie was somewhat of a cross between a canard aircraft and a biplane. The top wing canted slightly upward and the bottom wing canted slightly downward, making it also look like an X-wing starfighter head-on. *Courtesy of Burt Rutan*

Defiant

"I wasn't planning on kitting the Defiant when I built it. I built it for me," Rutan said, similarly to the intent of the VariViggen. "I wanted a twin. I wanted something that I would be comfortable in at night. And I wanted big baggage. I wanted a four-place," Rutan said.

For long-range, night, and bad-weather flying, twin-engine aircraft have an obvious advantage in the event of an engine failure compared to single-engine aircraft. However, twins can still get into trouble if one of the two engines fails. In this situation, the pilot must immediately compensate for the huge forces that develop from the asymmetry of having one engine generating thrust on one wing and the other engine generating drag on the other wing. This is especially dangerous on takeoff.

Rutan wanted a workhorse, so he took advantage of twin engines in a push-pull configuration. In RAF's newsletter, *Canard Pusher*, he described his "no-procedure-for-engine failure" design goal for the Defiant: "It doesn't take a lot of study to realize the impact on flight safety of a twin that not only has no appreciable trim change at engine failure, but requires no pilot action when it does fail. You can fail an engine at rotation for takeoff or during a go around in the landing flare. The pilot does nothing; he climbs out as if nothing happened. He has no prop controls to identify and feather. He has no cowl flaps to open, no wing flaps to raise, no min control speed to monitor (he can climb better than the other light twins even if he slows to the stall speed), no retrimming is required, he can even leave the gear down with only a 50 fpm climb penalty. The only single-engine procedures are the long term ones: (1) cross feed if you want to use all fuel on operative engine, (2) magnetos off."

Encouraged by the performance, Rutan decided to type-certify an aircraft based on the Defiant in 1979. To this point, all of his designs were for homebuilders. Now he hoped a pilot could go out and buy a completed Defiant instead of buying a twin from Beech, Cessna, or Piper. However, Rutan could not secure funding for the program. Since there was no shortage of projects running at RAF, Rutan shifted his focus away from developing the Defiant further. He still amassed more piloting time flying this aircraft than he did flying any other type of aircraft.

When the canard really gets too big, as with the tandem-wing Defiant, it is more of a wing. But it still provides stall resistance. And by building the twin-engine airplane with push-pull engines, Burt Rutan was able to make a much safer flying aircraft during a single-engine failure compared to other twins. *Courtesy of Burt Rutan*

As if a airplane made of composites with a canard, sweptback wings, winglets, and a pusher engine wasn't enough to excite crowds at Oshkosh in 1979, the Defiant capped it off with an engine in its nose as well. Note its winglet rising above the people and behind it, the winglets of the prototype Long-EZ, which got a major revision before plans were sold. *Courtesy of Burt Rutan*

Burt Rutan enlisted the help of expert homebuilder Fred Keller to take charge of developing a kit plan for the Defiant. Shown in front of the original Defiant (Model 40), Keller's Defiant homebuilt (Model 74) had larger wing-root strakes, wingspan, and winglets. *Courtesy of Burt Rutan*

Defiant Details

Model number	40
Type	twin-engine, push-pull, tandem-wing
Prototype tail number	N78RA
Current prototype location	Hiller Aviation Museum, San Carlos, CA
Customer	homebuilders, marketed 1984
Fabrication	RAF
Flight testing	RAF
First flight date	30 June 1978
First flight pilot	Burt Rutan
Seating	four-place
Wingspan	29.2 ft
Wing area	127.3 ft^2
Empty weight	1,585 lbs
Gross weight	2,900 lbs
Engines	two Lycoming O-320s, 160 hp each
Landing gear	tricycle, retractable nose gear & fixed main gear
Fuel capacity	90.5 gal
Rate of climb	1,750 fpm (gross weight)
Cruise speed	217 mph (65% power)
Range	1,290 miles
Ceiling	28,350 ft

The Homebuilt Defiant

In 1981, Rutan approached Alaskan Fred Keller, a grand champion winner at Oshkosh for the VariEze he built, to construct an improved version of the Defiant for homebuilders. Having a twin-engine in the rugged country of Alaska, not to mention developing a kit for Rutan, had very strong appeal for Keller.

Keller meticulously documented the building process. He would also be providing building support for the Defiant homebuilt.

Both wings now had greater spans, and the strakes at the rear wing roots were enlarged. Changes to the winglets and ailerons improved the Defiant homebuilt's flying qualities and performance. Modifications to the cockpit also afforded more room and comfort to the occupants. Like the prototype, it had two independent electrical systems and two independent fuel systems. In 1983, the cost for a homebuilder to build it, without avionics, was twenty thousand to thirty thousand dollars, requiring approximately 2,000 to 2,500 hours of labor for the construction.

The homebuilt Defiant, Model 74, tail number N39199, made its maiden flight on 16 July 1983, flown by Rutan and Keller. During this flight, however, Rutan noticed it wasn't climbing as fast as it should. The carburetor float stuck for one of the engines. They hadn't even realized they were flying along on just a single engine. The homebuilt Defiant's design proved its safety for an engine failure on the very first flight.

"The plans were put out in '84, and we closed up plan sales in '85," Rutan said. "So the Defiant plans were only sold for one year. Something like one hundred sets of plans in one year. And that's all."

Only a handful of this amazing aircraft ever got off the ground.

Long-EZ

Not too long after the VariEze homebuilt hit the market in 1976, Burt Rutan recognized the huge benefit of having much greater range for the aircraft. He began sketching out ideas to take advantage of a heavier and more powerful Lycoming engine. He initially called the new design the SuperEze.

"I showed that, hey, I can make an airplane with almost the same cruise speed, but it could go a couple thousand miles," Rutan said. "It would be a real safety thing to have, go around weather and whatever. And I was having trouble with people getting Continental engines. We were running out of them. Everyone had to buy a used engine."

When the long-range Longhorn Learjet came out, the first Learjet with winglets, Rutan liked the name, so he thought Long-EZ would be a great name for an aircraft with a range of around two thousand miles.

"I was kind of enamored by having something with fifty gallons of gas, and you just fill 'er up and go anywhere," Rutan said.

He wanted to design an airplane with more forgiving, easier to fly, and safer flying qualities. He even simplified the flight controls by using a small rudder on the nose

The last of the "EZ" trilogy, the Long-EZ had a cruising speed of 184 miles per hour, which was 9 miles per hour slower than that of the cruising speed of the VariEze homebuilt, but the Long-EZ more than made up for it by having a range of 1,970 miles, which was nearly double the range of the VariEze homebuilt. *Courtesy of Burt Rutan*

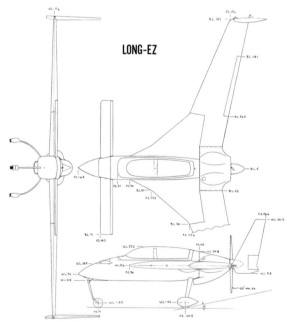

LONG-EZ

A continued refinement of a now familiar configuration, the design of the Long-EZ provided better handling for a more docile flying aircraft. Bigger overall, the Long-EZ's length, wingspan, and gross weight compared to the VariEze homebuilt were greater by 2.6 feet, 3.9 feet, and 275 pounds, respectively. *Courtesy of Burt Rutan*

called a rhino rudder. The pilot could see on either side, but if deflected, it obstructed the view.

"Now it had real big fuel strakes. That's where the fifty gallons came from," Rutan said. "And I scaled up the fuselage. Anyway, it was VariEze wings, VariEze canard, and made a little bit better."

In 1979, RAF flew the prototype to Oshkosh. The aircraft was far from having the design locked down, though. It was a terrible flying aircraft, and Rutan was very disappointed. He knew it wasn't something he could kit.

"I still told everybody, I said, 'Listen, I'm going to make this thing fly better. Make it fly really good. And I'm going to have plans for a long-range airplane.' That promise was made at Oshkosh '79," Rutan said. "By Oshkosh '80, we had the Long-EZ there at Oshkosh, and we had plans for it."

RAF's Most Successful Homebuilt

Dozens of modifications had been evaluated as the design refined step by step. Compared to the smaller VariEze, the Long-EZ had less sweep in the wings as well as less wing loading, which is just the aircraft weight divided by its wing area. Improved fabrication methods developed on the Defiant allowed for a 41 percent increase in wing area with only a small corresponding increase in weight.

The Long-EZ had slower landing speeds than the VariEze. Coming in on final approach was flatter, so the pilot didn't have any trouble seeing over the nose during landing. The stall speed was lower, too. The large winglets provided better directional stability, and the rhino rudder on the nose was abandoned in favor of winglet rudders.

"It was just safer and had more docile flying qualities," Rutan said. "It wouldn't go as fast for the same horsepower. It was a bigger airplane."

On 15 December 1979, Dick Rutan set a closed course world record, flying the Long-EZ, with an extra fuel tank in the back seat area, 4,800 miles nonstop. He flew for 33 hours and 34 minutes, averaging 145.7 miles per hour and 35 miles per gallon. On a separate, lightweight flight, he reached an altitude of 26,900 feet.

The typical Long-EZ pilot, though, could expect an economy cruise flying at twelve thousand feet, 144 miles per hour, and 3.52 gallons per hour. Stall-proof, the aircraft would climb at nine hundred feet per minute while at full aft stick.

"The Long-EZ is probably the best homebuilt that we did," Rutan said. "And I think it has probably the largest numbers of my designs that have been built."

First flown in 1979, the Long-EZ became Burt Rutan's best-selling homebuilt design. Having dialed in the balance between high performance and flying qualities, the Long-EZ gave a more comfortable ride over a longer distance compared to its "EZ" predecessors. This pair of Long-EZ's fly in tight formation amidst the low clouds. *Courtesy of Burt Rutan*

Long-EZ Details

Model number	61
Type	single-engine, canard pusher
Prototype tail number	N79RA
Current prototype location	prototype disassembled
Customer	homebuilders, marketed 1980
Fabrication	RAF
Flight testing	RAF
First flight date	13 June 1979
First flight pilot	Dick Rutan
Seating	two-place, tandem
Wingspan	26.1 ft
Wing area	94.8 ft² (including canard)
Length	16.8 ft
Height	7.9 ft
Empty weight	750 lbs
Gross weight	1,325 lbs
Engine	Lycoming O-235, 108 hp
Landing gear	tricycle, retractable nose gear & fixed main gear
Fuel capacity	52 gal
Takeoff distance	830 ft (gross weight)
Landing distance	680 ft (gross weight)
Rate of climb	1,350 fpm (gross weight)
Cruise speed	184 mph (75% power)
Range	1,970 miles (gross weight, 40% power)
Ceiling	22,000 ft (gross weight)

AD-1

In the early 1970s, the oblique or skew wing aircraft concept that NASA researcher Robert Jones invented in the mid-1940s got a good, hard second look. Jones had envisioned an aircraft that could pivot its wing, like the opening and closing of a pair of scissors, to make dramatic changes in flight performance. His calculations showed an aircraft would use half the fuel at one thousand miles per hour with the wing skewed at 60 degrees compared to an unskewed wing.

Now NASA looked to apply this concept to a high-speed airliner capable of flying Mach 1.4. The entire wing would start off perpendicular to the fuselage, like conventional aircraft, for low-speed flight during taking off and landing. For high speed, the entire wing would pivot around a connection point on top of the fuselage, angling one wingtip forward and the other wingtip backward. In the skew position, the drag would decrease, but both speed and range would increase.

Major aerospace companies wanted in, but NASA couldn't fund a very expensive research program. In 1975, two of Burt Rutan's college friends—one worked at Edwards Air Force Base and the other worked at NASA Dryden—approached him with an idea. "We've noticed that you built a VariEze in three months. Why don't you design for us an airplane with a skew wing?" Rutan said about them approaching him. "They noticed that I also did the jet version of the BD-5."

Rutan was still flight testing the VariEze at the time and hadn't yet started design on the Defiant. But he partnered up with Herb Iversen at Ames Industrial Corporation of Long Island, New York, the providers of the jet engine used in the BD-5J. NASA Ames and NASA Dryden also partnered for the research project.

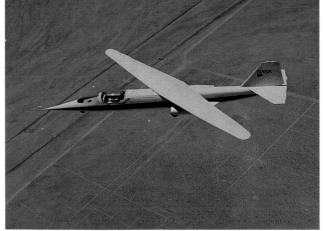

Above: The AD-1 was a technology demonstrator of a concept for NASA Ames and NASA Dryden. The concept was to have the wing perpendicular to the fuselage, like a normal airplane, to give the aircraft the lift needed for takeoffs and landings. However, in flight it would rotate the entire wing, skewing it to reduce drag. In this asymmetric configuration, cruise speed would be significantly improved. *Courtesy of Burt Rutan*

Left: Burt Rutan and Ames Industrial of Long Island, New York, teamed together to build the AD-1 for NASA. At 15 percent scale, the single-person aircraft was a technology demonstrator for a much larger futuristic, supersonic transport jet. *NASA*

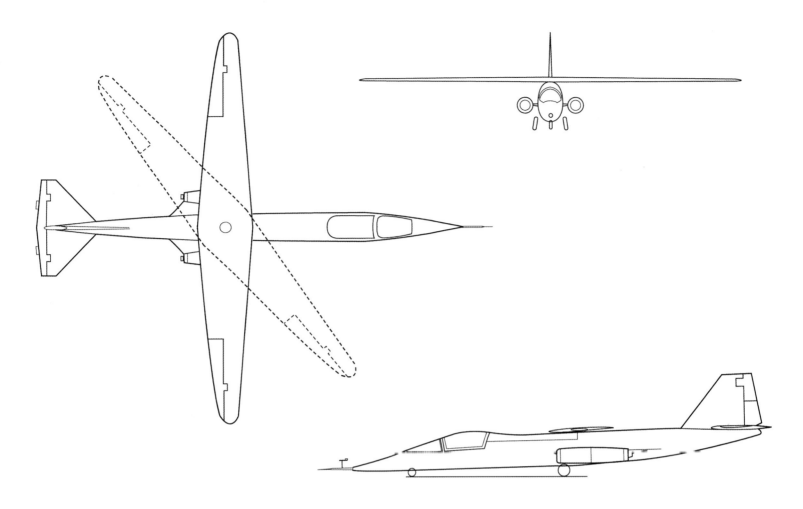

The flight characteristics of the oblique-wing, or scissor-wing, of the AD-1 were evaluated over a range of wing angles. An electric motor in the aircraft would rotate the entire wing in flight, as a single piece, from a position perpendicular to the fuselage, 0 degrees, to a maximum of 60 degrees. *NASA*

The Ames-Dryden-1 (AD-1), or Rutan's Model 35, was a 15 percent–scale version of a Boeing transonic airliner concept. Built using the same fiberglass and foam composite method as the VariEze and designed for study only at low airspeeds, the AD-1 had an electrically actuated wing that pivoted to a maximum skew of 60 degrees.

After seventy-nine flights total, the AD-1 completed all its technical objectives. It exhibited satisfactory flight characteristics at skew angles of 0 degrees and 30 degrees, but, as expected, the flying qualities declined from 30 degrees to 45 degrees due to aeroelasticity and pitch-trim coupling. From 45 degrees to 60 degrees of skew, the handling was poor, but the pilot could still compensate for the asymmetries. The AD-1 did not have flight control augmentation or mixing, which would have been used to compensate for these effects.

AD-1 Details

Model number	35
Type	twin-jet, skew wing research aircraft
Prototype tail number	N805NA
Current prototype location	Hiller Aviation Museum, San Carlos, CA
Customer	NASA
Fabrication	Ames Industrial Corporation
Flight testing	NASA
First flight date	21 December 1979
First flight pilot	Thomas McMurtry
Seating	single-place
Wingspan	32.3 ft (unswept)
Wing area	93 ft²
Length	38.8 ft
Height	6.75 ft
Empty weight	1,450 lbs
Gross weight	2,145 lbs
Engines	two Microturbo TRS18-046 turbojet engines, 220 lbs thrust each
Landing gear	tricycle, fixed
Fuel capacity	80 gal
Maximum speed	200 mph
Ceiling	12,000 ft

Danny Mortensen flew into Mojave one day and asked Burt Rutan to design him a very fast biplane for air racing. The result was the AMSOIL Biplane Racer. The struts between the wings, as shown, were needed to conform to the racing rules for biplanes, but it was normally flown without them. *Courtesy of Burt Rutan*

AMSOIL
Biplane Racer

Danny Mortensen stopped to refuel at Mojave on his way back from the Cleveland Air Races in 1979. He consistently took fourth or fifth place in his Mong biplane, and he felt that he needed a boost.

"I walked up to the door of the Rutan Aircraft Factory, knocked on the door, and went in," Mortensen said. "Burt was sitting there working, and I said, 'Hey, I want to go faster at the air races. You want to design an airplane for me? What's it going to cost?' And he said, 'Send me a copy of the biplane rules, and I'll take a look at them. I always wanted to go air racing.' "

Mortensen initially had the idea when he first talked to Rutan about modifying a Quickie to fit a Lycoming O-320 engine. Rutan then suggested building a whole new airplane instead.

"He designed three airplanes, and we took the least risky design," Mortensen said.

One of the radical designs, the Model 69, was a joined-wing aircraft, like a biplane where the wings slanted together to meet at the wingtips, sort of a reverse X-wing. Mortensen's choice, the original configuration of Model 68, was the easiest and quickest to build.

The AMSOIL Biplane Racer still looked like a Quickie, but it was substantially bigger, heavier, and faster. The aircraft used three different composites, Kevlar in the firewall between the engine and the cockpit, carbon fiber in the airfoils, and fiberglass for just about everything else.

Mortensen set a world speed record at 232 miles per hour, more than 100 miles per hour faster than a Quickie can fly. He had received his wish. It was so aerodynamically clean that at 150 miles per hour, it burned 7.5 gallons of fuel per hour.

"It had excellent visibility," Mortensen said. "When you went into a 70 degrees or 80 degrees bank around the pylons, you could see the entire racecourse and everybody out there. Whereas with the other biplanes, you had a wing over the top of the cockpit with a center pylon, and it blocked out your view. So you're kind of guessing in the dark."

Rutan's composite method was about to be put to the ultimate test, though. In 1983 at the Reno Air Races, a racer crossed in front of the AMSOIL Biplane Racer, and Mortensen found himself caught in wake turbulence.

"I got rolled almost inverted and didn't have enough control response to recover the aircraft in just three seconds, from one hundred feet, before hitting the ground," Mortensen said.

The aircraft knife-edged into the desert at two hundred miles per hour and then tumbled to a stop. Mortensen got up and walked away from the crash.

"Burt had said, 'I'm going to design this cockpit for 22 g in case something goes wrong out there on the racecourse.' He did a good job," Mortensen said.

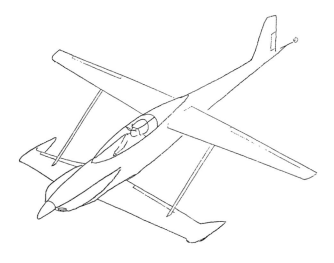

Figure 1   MODEL 68 ORIG. CONFIG.

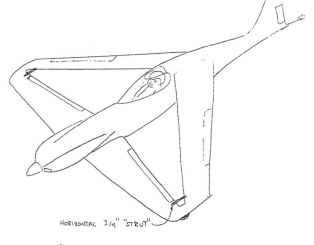

HORIZONTAL 3/4" "STRUT"

Figure 2.   MODEL 69

Burt Rutan came up with three concepts for Danny Mortensen to choose from. The most radical concept was a joined-wing racer. One concept had a wing mounted on the engine nacelle. Mortensen chose the third, which was the least radical of all the radical concepts. *Courtesy of Danny Mortensen*

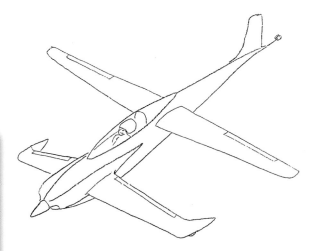

Figure 3.   ORIGINAL MODEL 68 WITH NACELLE WING ROOT.

AMSOIL Biplane Racer Details

Model number	68
Type	single-engine, biplane
Prototype tail number	N301LS
Current prototype location	EAA AirVenture Museum, Oshkosh, WI
Customer	Danny Mortensen/AMSOIL
Fabrication	customer
Flight testing	customer
First flight date	3 August 1981
First flight pilot	Danny Mortensen
Seating	single-place
Wingspan	22 ft
Wing area	79.6 ft²
Aspect ratio	6.36
Length	22 ft
Height	4 ft
Empty weight	854 lbs
Gross weight	1,167 lbs
Engine	Lycoming IO-320 A2B, 160+ hp
Landing gear	conventional, fixed
Fuel capacity	37 gal
Takeoff distance	750 ft
Landing distance	1,500 ft
Rate of climb	1,500 fpm
Maximum speed	232 mph
Cruise speed	170 mph
Range	600 miles
Ceiling	22,000+ ft

Though it looks like a Quickie, the AMSOIL Biplane Racer was substantially bigger. It was built with high-performance materials, including carbon fiber and Kevlar, whereas the Quickie was built only with fiberglass. And the engine of the AMSOIL Biplane Racer had nothing to do with economical flying—it had to do with power and speed. *Courtesy of Burt Rutan*

Next Generation Trainer (NGT)

"Years after we built the AD-1, but only shortly after NASA finished flight tests, along came Fairchild Republic," Rutan said. "And they figured out after looking at what we had done with the AD-1, that maybe we should build a flight test demonstrator of their design for the new air force trainer under the NGT program."

The Next Generation Trainer (NGT) was a replacement for the aging Cessna T-37 Tweet. The Fairchild Republic proposal called for a highly fuel efficient, twin turbofan two-seater with an H-tail, which was an unusual tail configuration at the time. So Fairchild Republic wanted to beef up its contract bid with some real hard data.

"There were three or four companies bidding on the NGT. They had low-speed wind tunnel tests," Rutan said. "But nobody thought that they could have manned flight tests with their proposal, particularly ones in which you could do departure, stall, and spin recovery tests.

This would really answer the customers' concerns about the configuration in a big way."

Model 73, a 62 percent–scale version of Fairchild Republic's Next Generation Trainer, incorporated fiberglass and foam construction with carbon fiber skins and spars. Ames Industrial Corporation would again supply BD-5J engines and fabricate the NGT as it had done for AD-1, which had been the first aircraft the company built from composites. Design and construction of the scaled-down NGT took eight months to complete, and during the following eight weeks, RAF conducted a rigorous flight test program.

The flight test data RAF produced was included with Fairchild Republic's bid. The air force awarded the contract to Fairchild Republic in July 1982. However, the NGT program fell apart when it was determined that the navy needed a new trainer as well. The eventual solution was to have a joint trainer instead of two separate trainers.

In order to win a contract for a new USAF trainer, Fairchild Republic hired Burt Rutan to build a 62 percent–scale technology demonstrator. By going subscale, it could be built extremely cheaply and quickly in order to provide flight test data that would help Fairchild Republic support its bid. *Courtesy of Burt Rutan*

The NGT was powered by twin Microturbo jet engines provided by Ames Industrial. The full-scale NGT aircraft would run on larger, higher performing jet engines. One of the key elements evaluated during the flight testing was the unconventional H-tail of the NGT. *Courtesy of Burt Rutan*

Next Generation Trainer Details

Model number	73
Type	twin-jet, subscale demonstrator
Prototype tail number	N73RA
Customer	Fairchild Republic
Fabrication	Ames Industrial Corporation
Flight testing	RAF
First flight date	10 September 1981
First flight pilot	Dick Rutan
Seating	single-place
Wingspan	21.9 ft
Length	17.8 ft
Height	5.7 ft
Empty weight	900 lbs
Gross weight	1,500 lbs
Engines	two Microturbo TRS18-046 turbojet engines, 220 lbs thrust each
Landing gear	tricycle, retractable
Maximum speed	288 mph (never exceed speed)
Cruise speed	250 mph

Because it was subscale, that meant everything was subscale except for the size of the pilot. So even though it was a side-by-side trainer, the subscale NGT could only fit one pilot. *Courtesy of Burt Rutan*

Grizzly

"The Grizzly is an airplane built for camping and landing in meadows," Burt Rutan said. "It had room for sleeping in the back end. I did it because I was getting some critiques that all these canard airplanes have real nice performance, but they don't have very good slow, short-field performance. So I thought, well, I'll just do one that has short-field performance."

At a landing speed of seventy-four miles per hour, the VariEze came in quite a bit faster and shallower than most light airplanes. The Long-EZ design improved upon this, but not by much.

Rutan quietly designed Grizzly, which he didn't intend to market but did intend to prove that he was up to the challenge.

Grizzly deployed, as shown, huge Fowler flaps on both the canard and main wing. The canard Fowler flaps were a record 55 percent of the canard's chord. When engaged, they extended backward, increasing wing area, and downward, increasing the curvature of the wing. These changes in the shape gave the aircraft better lift, reduced the stall speed, and increased drag, allowing for slower landing approach speeds. *Courtesy of Burt Rutan*

A research aircraft, Grizzly had a right wing with a standard skinned foam core construction, and its left wing tested a new honeycomb sandwich structure. Specialized for short takeoff and landing (STOL), it could be used for backcountry flying and camping. Grizzly had long, curved front landing gear with doubled-up tires suitable for rough terrain. *Courtesy of Burt Rutan*

Made from fiberglass and foam construction with carbon fiber strengthening, Grizzly had a forward-swept canard and a main wing that were attached together on each side by a stiffening boom, which ran from the canard tip to midspan on the main wing. Only the main wing inboard of the boom was forward swept. Ailerons were mounted on the outboard sections of the main wing, and the tail had a full flying horizontal stabilizer. To make room for very large Fowler flaps, the booms carried the fuel.

This was the first aircraft to be flown by Rutan using forward-swept wings. Also, Fowler flaps were typically used on much heavier and bigger airplanes.

Anyone sitting by a wing of an airliner as it's getting ready to land will see flaps engage where the back edge of the wing extends outward and downward. A big increase in lift results, enabling the airliner to approach at a steeper angle and to land at a slower speed. Thus, it stops in a shorter distance once its wheels hit the runway.

The pilot also gets a better view since the nose does not pitch up as high since flaps create a large amount of drag.

When Grizzly's Fowler flaps extended, they increased the canard and wing widths by a whopping 55 percent. This added forty-five square feet to the wing area.

"It had great takeoff and landing distances, and a lift coefficient of about 3.7—twice a normal light plane," Rutan said.

Ultimately, Grizzly was a research project. It proved to have excellent short takeoff and landing (STOL) capabilities. Rutan had fun working on Grizzly and learned about new flying characteristics, mechanical systems, and construction methods.

He had also wanted to investigate using Grizzly as a seaplane, where it could easily move back and forth between water and land. However, with a new homebuilt sailplane in development and other projects running full steam ahead, he had to choose to put Grizzly aside.

Grizzly Details

Model number	72
Type	three-surface, STOL
Prototype tail number	N80RA
Current prototype location	EAA AirVenture Museum, Oshkosh, WI
Customer	RAF R&D
Fabrication	RAF
Flight testing	RAF
First flight date	22 January 1982
First flight pilot	Mike Melvill
Seating	four-place
Empty weight	1,474 lbs
Gross weight	2,494 lbs
Engine	Lycoming IO-360B, 180 hp
Landing gear	conventional, heavy duty, fixed
Minimum speed	40 mph
Cruise speed	132 mph

Most bush planes have high wings to ensure they clear the bush and to give them a better view of the ground below. Because this type of airplane flies so close to such irregular terrain, it is important for the pilot to see the condition of the ground before landing on what couldn't be expected as a smooth surface. Grizzly had low wings, but it had bubbled-out side windows that allowed the pilot and passengers to see straight down between the canard and main wing. *Courtesy of Burt Rutan*

Solitaire

During the time of Grizzly's construction, the Rutan Aircraft Factory was an open shop. So when Burt Rutan decided to build Grizzly under wraps, it was hard to hide the fact that he was working on a new aircraft design.

Since he wasn't going to sell plans for Grizzly, he didn't want homebuilders to stop buying Long-EZ plans, thinking there was a new generation on the way when there wasn't. The Soaring Society of America's self-propelled sailplane competition scheduled for 1982 in the nearby mountains of Tehachapi gave Rutan an idea. He had always wanted to build a sailplane. After all, the very basis of the composite construction methods he pioneered originated from sailplane repair methods.

So by announcing that he would enter the competition, he was able to keep people's focus on the hand that built Solitaire while other built Grizzly in secret.

Typically, a sailplane requires a tow plane to pull it up into the air, but a self-propelled sailplane has its own small engine to do this. The advantage is obviously that the pilot does not have to wait around for a tow plane. The disadvantage is that it has to incorporate all the extra weight and complexity of an engine, so it will be heavier and have worse performance than a sailplane without an engine.

Solitaire's entire engine folds out on a pylon from a compartment in front of the pilot. After Solitaire took off and ascended, it retracted the engine for gliding. Both Rutan and Mike Melvill, the test pilot, were able to fly

At the time Solitaire was one of the most advanced composite kits that homebuilders could purchase. Its preformed fuselage had a honeycomb sandwich structure of fiberglass and Nomex. In this static force test, the wingtip bent down 45.5 inches, but once the load was removed, the wing returned back to its normal shape. The aircraft could handle +7/-3 g. *Courtesy of Burt Rutan*

Shown with its engine stowed, the self-launching sailplane Solitaire (Model 77) had airfoils that were specially designed by John Roncz to provide maximum soaring performance. Solitaire had a glide ratio of 32:1, minimum sink rate of 150 feet per second, and maximum glide speed of 132 miles per hour. *Courtesy of Burt Rutan*

The entire 23-horsepower KFM 107E engine rose out from a compartment in front of the pilot. By using this engine, Solitaire didn't need the help of a tow plane. Construction was estimated to take 400 hours and cost between $7,000 and $9,000, which included the engine. *Courtesy of Burt Rutan*

Solitaire Details

Model number	77
Type	self-launching sailplane
Prototype tail number	N81RA
Current prototype location	Reedley College, Reedley, CA
Customer	homebuilders, marketed 1983
Fabrication	RAF
Flight testing	RAF
First flight date	28 May 1982
First flight pilot	Mike Melvill
Seating	single-place
Wingspan	41.8 ft
Wing area	102.4 ft^2 (canard and main wing)
Aspect ratio	20.79
Length	19.2 ft
Height	5.3 ft
Empty weight	380 lbs
Gross weight	620 lbs
Engine	KFM 107E, 23 hp (retractable)
Landing gear	tandem nosewheel and mainwheel, wingtip wheels, fixed
Fuel capacity	5 gal
Max L/D	32
Takeoff distance	940 ft (powered)
Landing distance	300 ft (powered)
Rate of climb	500 fpm (powered)
Maximum speed	81 mph (powered)
Cruise speed	63 mph (powered)
Range	150 miles (powered)

longer than two hours without having to use the engine.

Rutan used a hammerhead-style canard design for Solitaire, where the canard's leading edge was flush with the tip of the nose. He felt that this configuration helped reduce drag. Sailplanes must be very maneuverable and be able to alter their glidepath in order to catch thermals, the rising air that helps lift them up like soaring birds.

Normally, sailplanes use spoilers and flaps to make these adjustments. However, these control surfaces would cause dramatic pitching of Solitaire's nose up or down because of resulting changes to the lift of the wing.

Rutan had to design what's called a spoilflap, which created drag but did not change the lift of the wing, so Solitaire could change speed without pitching its nose.

Since Rutan had finished his research with Grizzly by now, he then got it qualified as a tow plane. So during flight testing, or when time was an important factor, Grizzly could get Solitaire high into the air much faster. In the end, Solitaire helped get Grizzly built, and then Grizzly helped Solitaire get built.

Rutan, a newly qualified tow pilot, with Melvill in tow in Solitaire, took off from Mojave when the time came for the competition. But Rutan released Melvill far enough away from the airport so that when he landed, Melvill was nowhere to be seen. That was until Melvill snuck up on the airport and did three loops over the runway. Solitaire won the competition as well as later receiving an outstanding design award from the Soaring Society of America.

Solitaire was far from a commercial success, though. "Cost me twice as much to develop this as the Long-EZ, and I sold about twelve kits," Rutan said.

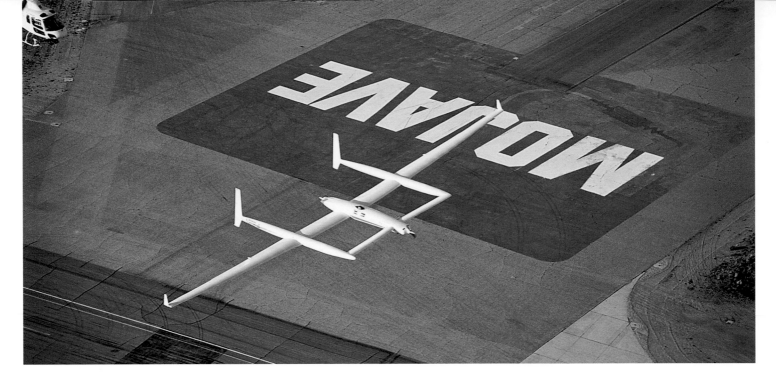

Voyager

In the late 1960s, Jim Bede built the BD-2 *Love One* (Low Orbit Very Efficient), which was based off a conventional sailplane design, to fly around the world. In the early 1980s, the Quickie Aircraft Corporation began development on the world flyer *Big Bird*, also a monoplane based off a sailplane. Tom Jewett of QAC had worked with Burt Rutan for Bede.

"In my opinion, neither one of those airplanes had enough range to fly around the world," Rutan said. "So I had kind of in the back of my mind that someday I may try it."

Quite a rivalry had developed between RAF and QAC by this time. Rutan started to make calculations himself about the feasibility of a round-the-world (RTW) flight. He even considered a flying wing. A B-52 bomber held the long-distance record at the time of 12,532 miles, which amounted to only halfway around the planet.

That was about it until Dick Rutan and his partner Jeana Yeager called a meeting with Burt Rutan in 1981 at a cafe to discuss building a world-class aerobatic airplane called Monarch.

By this point, Burt Rutan was very busy with the design of Starship, a large, twin pusher engine, business aircraft. However, he had a policy of not doing high-risk airplanes for the public, so he offered a counterproposal.

"Dick had already set a world distance record in my Long-EZ. And he was just drooling at the mouth to do world records," Burt Rutan said. "And I mentioned that it is possible for what Bede tried to do. 'I think I have a design that will do that. Why don't you do that instead of this aerobatic thing?' "

They wholeheartedly agreed. So he revealed to them his concept of what would become Voyager. "I drew on a napkin a sketch of what I had been designing. So that

Voyager was designed for only one specific task—to fly around the world, nonstop, and without refueling. The closest attempt made by any other aircraft beforehand was in a B-52, a long-range strategic bomber with eight jet engines. Halfway around was only as far as the B-52 could get, though. A monumental challenge awaited Voyager. *Courtesy of Burt Rutan*

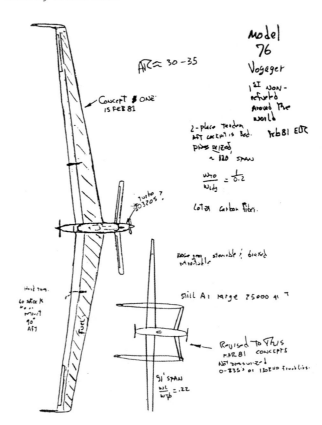

Burt Rutan had judged that *Love One* and *Big Bird*, two aircraft designed for a round-the-world flight, did not have the adequate performance to make it all the way. So he began to think about the feasibility of building an aircraft that could do it. This early sketch, drawn by Rutan, shows two designs he considered. Burt Rutan would soon after partner with Dick Rutan and Jeana Yeager to design, build, and fly Voyager. *Courtesy of Burt Rutan*

napkin was actually not an original idea. That napkin was a conveying of the configuration to Dick and Jeana, who had not seen it before."

The design centered on the single most important element required to circumnavigate the globe. Just as with the *Spirit of St. Louis*, Voyager was a flying fuel tank—seventeen fuel tanks to be exact. Because of the dramatic difference in weight of Voyager at the beginning a flight compared to the end as fuel burned off, Burt Rutan decided that by staging two state-of-the-art engines, he could keep one or two of the engines running at the highest efficiency possible.

The engines had to be inline, push and pull, because as the flight progressed, one of the engines would be shut off. The wing had to be designed for soaring, so it was long and thin like Solitaire. However, since it would be laden with fuel, there needed to be additional support. Rutan added a canard but still needed vertical stabilizers and rudder control.

Where to put more fuel as well? The fuselage had the cockpit and engines. So by connecting the canard to the wing with booms, like Grizzly, he could solve how to support the wing and where to put the tail and fuel.

Carbon fiber honeycomb sandwich structure formed the shell of Voyager, and every open space inside was filled with fuel. The super-lightweight structure alone weighed an astonishingly low 938 pounds, which was slightly more than three times what the crew would weigh.

It took an outpouring of support, donations, and volunteer hours to get Voyager off the ground. But funding wasn't the only challenge.

"Voyager was built uncompromised for range at the expense of a whole bunch of other things—structures, flying quality, system reliability, all those things," said Dick Rutan, who flew the flight test program. Every flight had a problem.

The size of the right boom was increased slightly forward at the tip in order for it to house a weather radar. And to improve Voyager's performance, the wing was extended and winglets were added. The extensions on each side of the wing did not contain fuel but were foam core. Voyager had never been filled to the max with fuel before, and as Voyager took off on its round the world flight, its wingtips scraped on the runway. *Courtesy of Burt Rutan*

Pusher Engines are Pushed Aside

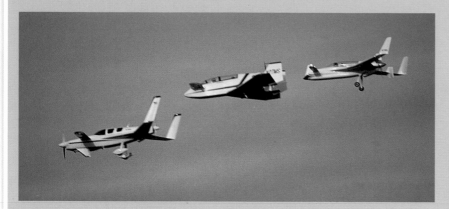

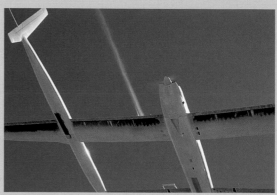

Burt Rutan has gone through a conversion in his understanding of pusher engines and came to realize some observations he made about them in the past aren't accurate. "I said that if you've got a pusher, it draws the air towards the propeller and keeps from having separation on the back of the fuselage. So it's actually less drag, and, of course, it's quieter because the engine and propeller are way back there instead of right here in front of you."

Back during the development of his pusher-engine designs, Rutan studied engine position using a Cessna 337 Skymaster, a push-pull twin. He made a performance comparison of the different positions by switching off one engine at a time while feathering the stopped engine's propeller, which turned the propeller blades parallel to the airstream to reduce drag.

"If you feather the front engine, go full power on the back, do a single-engine climb, and measure your climb performance," Rutan said of the comparison, "then feather the back engine, go full power of the front, and repeat the maneuver, the airplane climbs better if the back engine is running and the front is feathered."

Back then, he reasoned from this performance that pusher engines were more efficient than tractor engines. He now admits that he was wrong. "They aren't more efficient," he said. "It's just that the Cessna Skymaster had such a horribly blunt back end with massive separation and drag. By running that back propeller, it made that a lot better. This isn't true on a Defiant or a VariEze or a Long-EZ when you have a good cowling."

As far as the sound level, the interaction between the wing and pusher engine produced a significant source of noise. As the pusher propeller spins around, it encounters fast moving air as it swings above and below the wing. But each time it passes directly behind the wing, it hits slower moving air. So the propeller vibrates, shaking fore and aft, because during each full rotation, the blades go from fast air to slow air to fast air to slow air.

"Pushers will get shockwaves momentarily, twice per revolution. And that's why when you hear a pusher fly over, like an Avanti or a Starship or a Defiant or a Skymaster, you hear this bad fretting," Rutan said.

Voyager was a push-pull, requiring both engines for staging. Specially designed to accomplish a single mission, Voyager ran only one engine during most of round-the-world flight while the other engine feathered. Efficiency was critical, of course. But Rutan would soon consider designing smaller aircraft with the goals of being the most efficient and having the longest range.

"I was doing Catbird and Boomerang," Rutan said. "And there's no way I would use a pusher."

Above: Voyager required two engines because staging them was more fuel efficient than using a single engine and slowly reducing its power, thus its efficiency, as fuel burned off. But in order to keep symmetrical thrust when only one of the two engines ran during staging, they needed to have an in-line, push-pull configuration. *Courtesy of Burt Rutan*

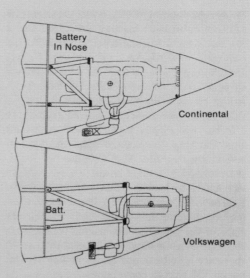

This drawing shows the installation of a 100-horsepower Continental aircraft engine and a 62-horsepower Volkswagen automotive engine in a VariEze homebuilt. Because the Continental was more powerful and massive, it took up the whole engine compartment so even the battery had to be moved forward to balance the weight. *Courtesy of Burt Rutan*

"I'd tell him how bad it flew, and Burt said, 'Remember our agreement.' And the agreement was that it had to have mission adequate flying qualities," Dick Rutan said. "So since it was a world flight, it really technically didn't have to turn at all. Maybe Burt told me that tongue-in-cheek. But if we had to fix any of those things that were wrong with it, we wouldn't have made it around the world. So actually in Burt's genius, he did it just right."

Voyager actually did have good flying qualities when it wasn't full of fuel. Dick Rutan praised its handling on the first flight. But Voyager would not be flying around the world with only two or three days' worth of fuel.

"When it was flown with about four days' fuel onboard, it became a lumbering beast that required a lot more pilot attention than when it was light, particularly in turbulent air," Burt Rutan said. "When it was loaded above a seven-day fuel load, it became what is known as *dynamically unstable in pitch*. This was a dangerous flying characteristic that could be damped and smoothly flown by the autopilot. However, when the pilot was flying it without autopilot, it required his complete attention to keep it flying safely. Inattention would result in the aircraft diverging in a wing-flapping mode such that it would destroy itself in less than a minute."

One of Voyager's winglets, damaged by scraping along the runway, fell off right after takeoff. To make Voyager symmetrical, Dick Rutan was able to snap off the other winglet using aerodynamic forces. After very careful inspection, no fuel was found to be leaking from either wingtip. Voyager then pressed on to begin its record-breaking journey. *Courtesy of Burt Rutan*

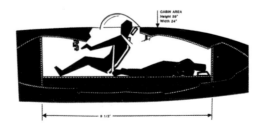

EAA's AirVenture Museum in Oshkosh, Wisconsin, has an expansive exhibit dedicated to Voyager, which includes a full-size replica of its fuselage with a cutaway of the cockpit. This photograph of an exhibit diagram shows the amount of space Dick Rutan and Jeana Yeager had in Voyager as they circled Earth. *Dan Linehan*

After setting a world closed-course distance record of 11,857 miles on a warmup flight doing laps up and down the West Coast five months earlier, Dick Rutan and Jeana Yeager climbed into the cockpit of Voyager to set off on their journey like the world explorers of old. On 14 December 1986, Rutan looked out through the small bubble canopy as the long runway of Edwards Air Force Base stretched out as far as he could see. This was nowhere near as seemingly endless as the potential thirteen days of nonstop, nonrefueled flying that waited past the end of the runway.

It took nearly three miles for Voyager, which was 72 percent fuel by weight, to lift off. Voyager flew west over the Pacific Ocean toward Hawaii to make use of the strong tradewinds and fly around the fattest part of Earth. The flight plan called for Voyager to cross over as much ocean as possible to avoid having to

A push-pull engine configuration propelled Voyager around the world. But both engines were not always on at the same time. As fuel burned off, less power was needed. Instead of reducing the power of both engines, it was more fuel efficient to run only one at higher power. *Courtesy of Burt Rutan*

After flying for 9 days, 3 minutes, and 44 seconds, covering a distance of 24,986 miles, evading a typhoon, dodging hostile aircraft, riding out thunderstorms, performing in-flight repairs, resuscitating a sickly engine, being cooped up together for so long in such a small space, landing with only 2 percent of the original fuel load, and smashing the previous world record, it was time for hugs. *Courtesy of Burt Rutan*

coordinate with foreign countries during the flyover. Dick Rutan and Jeana Yeager skirted a typhoon north of Papua New Guinea and tried to navigate around storms that seemed to trace from the tip of India to the coast of South America.

The crew tackled the loss of wingtips minutes after takeoff; an autopilot failure; oil, coolant, fuel problems; and an engine out on the homestretch with the coast of California just about in sight. But aside from weather and mechanical problems, Rutan and Yeager braved brutal fatigue cramped together in Voyager's tiny cockpit.

On December 23, Voyager touched down after circling Earth on a record-shattering 24,986-mile path while in the air for 9 days, 3 minutes, and 44 seconds. With an average speed of 116 miles per hour, Voyager had burned through all but 140 pounds of its starting 7,011.5 pounds of fuel.

Voyager Details

Model number	76
Type	long range, twin engine
Prototype tail number	N269VA
Current prototype location	National Air & Space Museum, Washington, DC
Customer	Voyager Aircraft Inc.
Fabrication	RAF
Flight testing	RAF, Voyager Aircraft Inc.
First flight date	22 June 1984
First flight pilot	Dick Rutan
Seating	2
Wingspan	110.8 ft (with winglets)
Wing area	362 ft²
Aspect ratio	33.8 (wing), 18.1 (canard)
Length	25.4 ft (fuselage), 29.2 ft (boom)
Height	10.3 ft
Empty weight	2,250 lbs
Gross weight	9,694.5 lbs
Engines	Teledyne Continental O-240, 130 hp (front)
	Teledyne Continental IOL-200, 110 hp (rear)
Landing gear	tricycle, retractable (manually)
Fuel capacity	7,011.5 pounds
Takeoff distance	14,200 ft (gross wt)
Maximum speed	150 mph
Speed for max range	135 mph (first day RTW)
	85 mph (last day RTW)
Range	24,986 miles (RTW)
Ceiling	21,000 ft

Burt Rutan designed Catbird, Model 81, with the sole purpose of being the most efficient light aircraft ever in order to compete in the CAFE 400 race and take the title away from RAF's flightline adversaries, the Quickie guys. In 1988, Catbird did claim the title, having scored more than 18 percent higher than any of the previous years' high scores. *Courtesy of Burt Rutan*

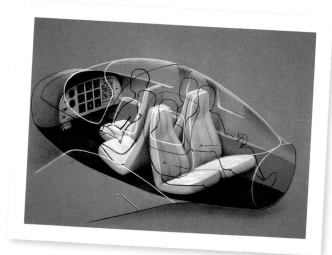

Even the seating of Catbird emphasized efficiency. The pilot and passengers were packed to minimize the required cabin space by considering the shape of people sitting down. So the pilot sat centered in front, giving the front row passengers leg room to the sides, and the back row passengers faced the tail with their legs in the narrowing part of fuselage.
Courtesy of Burt Rutan

Catbird

In 1981, efficiency aficionado Brian Seeley devised the 250-mile-long Competition in Aircraft Fuel Efficiency 250 (CAFE 250) and in the following year expended it to the 400-mile-long CAFE 400. RAF flew its VariEze, Long-EZ, and Defiant up to Santa Rosa, California, to compete in this annual race.

"It was a big knock down, drag out after the Quickie guys started their business," Burt Rutan said. "They had a Q2. And they went up there and beat us with this Q2 or Q200. So they would put out these ads in *Sport Aviation*, even in *Flying* I think—the world's most efficient airplane. That kinda burned me."

Rutan was part of the design team with Gene Sheehan and Tom Jewett that developed the Quickie (Model 54). The Q2 was a larger, two-place version of the Quickie, and the Q200 was a step up from the Q2, running with the more powerful Continental O-200 engine.

"So I decided to build an airplane they couldn't beat," Rutan said.

It wasn't an airplane that Rutan really needed. It was really the challenge he needed.

After building and flying thirteen different types of manned aircraft, Catbird came along looking very much like the typical aircraft Rutan seemed to do his very best to steer away from. Don't be fooled. Catbird's whiskers are actually a forward-swept canard, and the t-tail and wings are also forward swept.

"This is an interesting part of the RAF story because it was a significant event that we went to every year," Rutan said. "We planned for it. We tweaked our airplanes performance. It affected the work that we did to get better efficiency on airplanes, which usually you wouldn't pay a lot of attention to cause someone will buy a VariEze because of how looks, not what the fuel flow is."

Seeley understood that there was a limit on the efficiency of an airplane due to the laws of physics and properties of air and fuel. To measure efficiency for the race, he came up with a formula based on average speed, amount of fuel used, and mass of payload. The winner had to post the highest score using this formula.

"The airplane had to have a certain takeoff and landing performance," Rutan said. "He didn't say a stall speed. But he said you had to get off the ground on the takeoff on this four hundred–mile trip with all the people on board. You had to climb over this string on the runway."

Rutan felt that advertised "brochure performance" of aircraft did not always measure up to actual performance, so the CAFE 400 gave a good way to show off the performance differences between aircraft that flew governed by the same set of rules.

Race officials on mountaintops watched as the racers made their way around turning points. "To get miles per gallon, they would weigh the airplanes before and after the flight and watch everybody," Rutan said. "They had a certain allowance because people would sweat and perspire. You'd bring food along—four hundred miles was a long time. But then you'd eat it and it ended up in your bodies. They weighed the people separately. And they did all kinds of things to make sure you couldn't cheat. But by these two weighings, they'd get the amount of fuel that you used."

Catbird was optimized to reduce aerodynamic drag. The wetted area, or area that comes in direct contact with the airstream, was minimized by eliminating everything possible sticking off the fuselage, wings, and empennage. The overall shape looked like a sailplane.

"It had the very best airfoils in the world," Rutan said. He had John Roncz design these airfoils—the cross-sectional shapes of the wing—specifically for Catbird. Roncz had also designed airfoils for several other RAF airplanes, including Solitaire and Voyager.

Rutan's second company, Scaled Composites, had been in operation for several years. At the time of Catbird's development, it was owned by Beechcraft. Jim Walsh, the president of Beechcraft back then, got to see what Rutan was building.

"He envisioned that, 'Hey, the technology that's gonna come out of this in terms of performance is world class, and that is of interest to Beechcraft. Because if

we do a Bonanza replacement, we want to get the very best performance that you can get'—the best range, the best cruise speed, and so on. So he said, 'Listen, I'm interested in that. Why don't we help you pay for it?' So the Catbird for about a year or so got funding. The people who were building it, just a couple of people, got paid by Scaled, which was owned by Beechcraft. And they also bought the engine for it."

In 1988, Catbird, piloted by Mike Melvill, won the CAFE 400, having also set a new all-time high score by the wide margin of nearly 18 percent above the previous record. In 1994, Dick Rutan broke a world speed record in Catbird, flying a 1,243-mile-long (2,000 kilometers) closed course at 246.5 miles per hour. About a month later, Melvill set the same record, but with Catbird at a weight class one level higher, at 257.1 miles per hour. Catbird had accomplished the weight class jump by adjusting fuel and payload. These two world records currently stand.

Catbird now hangs upside-down in one of Scaled Composites' hangars. "It's a pretty airplane on top. Almost every airplane is like that," Rutan said. But why should such a successful aircraft be relegated to the rafters? The answer to this was the arrival of Rutan's favorite and arguably one of his more unconventional-looking designs.

Catbird now perches, upside down, from the rafters of a hangar in Scaled Composites. This location and orientation may seem to be a strange place for an aircraft that currently holds two world records. But the top planform is Burt Rutan's favorite view. And the Catbird's engine moved to his favorite airplane, the Boomerang. *Dan Linehan*

Catbird Details

Model number	81
Type	single-engine, three-surface
Prototype tail number	N187RA
Current prototype location	Scaled Composites (static display)
Customer	RAF R&D/Beechcraft
Fabrication	RAF/Scaled Composites
Flight testing	Scaled Composites
First flight date	14 January 1988
First flight pilot	Mike Melvill
Seating	five-place
Wingspan	32 ft
Wing area	100 ft^2
Aspect ratio	10.24
Empty weight	1,425 lbs
Gross weight	2,850 lbs
Engine	Lycoming TIO-360C1A6D, 210 hp
Landing gear	tricycle, retractable
Fuel capacity	74 gal
Maximum speed	272 mph (100% power)
Cruise speed	251 mph (75% power)
Cruise altitude	24,000 ft

The Boomerang is Burt Rutan's favorite aircraft. When he reflects about all his designs, he regards the Boomerang as his best achievement in general aviation because of its safety, performance, configuration, and the way it was fabricated. *Courtesy of Burt Rutan*

Boomerang

"Boomerang is a phenomenal airplane from the standpoint of its range, its performance, its speed, and its noise level. The fact that you do not even need to touch the rudder pedals in order to fly it at its stall speed with an engine out is something that is unheard of for multiengine airplanes," Burt Rutan said of the design of his that is the most widely misunderstood and underappreciated. "It is the airplane I am most proud of."

When Rutan wanted to fly anywhere, he had a small fleet to chose from, a VariEze, Long-EZ, Defiant, Catbird, and Grumman Tiger, which was the company's utility aircraft. His highest amount of flying time in any one aircraft came in the Defiant.

While the Boomerang's normal cruise speed is a blistering 263 knots (303 miles per hour), when slowed to its long-range speed of 186 knots (214 miles per hour) and an altitude of 19,000 feet, the Boomerang could comfortably carry four people a range of 2,600 nautical miles (2,990 statute miles), without the need of ferry tanks. That's plenty of distance to cover a flight from Mojave to Hawaii. *Courtesy of Burt Rutan*

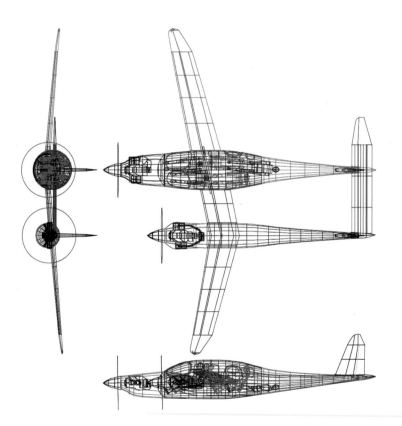

A twin-engine aircraft is significantly more of a challenge to design than is a single-engine aircraft. This is because a twin must be able to fly not just with two engines but also when one of the two has failed. So the design consideration goes from two permutations with a single—either on or off—to four permutations with a twin—either both on, engine #1 on and engine #2 off, engine #1 off and engine #2 on, or both off. *Courtesy of Burt Rutan*

"I felt more comfortable particularly at night and certainly over the mountains to have two engines," he said.

However, Rutan felt that he had one more homebuilt in him. In the early 1990s, under Rutan Designs, he started privately funding the development of Boomerang. Rutan occasionally does work that is not associated directly with RAF or Scaled Composites. Rutan Designs is a personal business that he put together for this purpose.

"It was the airplane that I built for myself so I could fly to Europe and Australia. I wanted to fly around the world. I wanted to do all kinds of things with the Boomerang. So Catbird wasn't all that attractive to me. Except for one thing, it had a TIO-360—a turbocharged, constant-speed prop engine—battery, and everything. So I moved it over to the Boomerang to save money."

Rutan laid out a set of high-performance specifications for himself that focused not just on speed but on range. Even more important than these criteria, though, was that he wanted a multiengine airplane with very safe engine-out characteristics. His push-pull Defiant fell into this category, but also fell short on what he now pursued.

"There's a bunch of things about a push-pull that surprised me on the Defiant performance-wise and noise-wise and vibration-wise," Rutan said. "My mind kept saying don't even go there. Don't try to solve those problems."

Rutan's previous three RAF designs included Solitaire, Voyager, and Catbird. All had sailplane-like qualities. He wanted this new twin to look and perform like a sailplane with its long slender fuselage and small tail.

"I just knew that to get a big jump in performance I had to do something that was kind of a clone halfway between a conventional light twin and a competition L/D 60 sailplane. And I kept coming back to this asymmetric configuration."

Rutan shaped and reshaped this aerodynamic puzzle of causes and effects as he tried to balance performance and safety. He was not satisfied that these were mutually exclusive.

He found common ground by shifting engines back and forth along the wing until ultimately one engine left the wing entirely. To offset weights, he shifted the wing more to one side and tilted and swept sections of it. An engine nacelle boom added structural strength and stowing capacity. Twin rudders trailed behind the engines directly in their propeller wash.

The center of gravity (CG) is a critical balance point for any aircraft. For a twin, the farther away the engines are from the CG, the more turning force results if one engine stops running. So a greater counteracting force is then needed to keep the aircraft on course. Boomerang's configuration dramatically reduced the distance of the engines from the CG, and its built-in aerodynamics naturally provided much of the counteracting force.

What Rutan created was an airplane that looked asymmetric but flew symmetrically. If either engine went out, Boomerang still flew symmetrically with very little required to compensate for the dead engine. No other twin-engine aircraft comes close to this.

When it came time to build Boomerang, Rutan developed a new manufacturing method to build the fuselage.

"I was going to build an airplane that was filament-wound carbon fiber and was grid stiffened, not sandwich. As part of the manufacturing process, I had this one-pound foam that was part of the tool, and then I just put a ply on the inside. I had an absolutely airtight and very well insulated cabin."

By making a male tool in the shape of the fuselage with the foam insulation layer grooved, he easily spun the carbon fiber around it. It took only fourteen hours to spin up the fuselage and several other major components.

Aside from the breakthroughs in asymmetric design and one-piece, filament-wound fuselage fabrication, Rutan had developed an idea for a revolutionary pressurization system and intended to come up with a state-of-the-art avionics system for Boomerang.

Even though too many other projects pulled him away from completing these final two goals, Rutan emphasized, "Boomerang is my best work in general aviation by a long margin."

Boomerang Details

Model number	202
Type	twin-engine, asymmetrical configuration
Prototype tail number	N24BT
Current prototype location	still flying
Customer	Rutan Designs
Fabrication	Rutan Designs
Flight testing	Rutan Designs
First flight date	19 June 1996
First flight crew	Mike Melvill (pilot) and Burt Rutan (copilot)
Seating	five-place
Wingspan	36.7 ft
Wing area	101.7 ft^2
Aspect ratio	13.2
Length	30.6 ft
Empty weight	2,370 lbs
Gross weight	4,242 lbs
Payload	813 lbs (with full fuel)
Engines	Lycoming TIO-360A1B, 200 hp (on boom)
	Lycoming TIO-360C1A6D, 210 hp (on fuselage)
Landing gear	tricycle, retractable
Fuel capacity	1,007 lbs
Takeoff distance	2,750 ft
Landing distance	2,580 ft
Rate of climb	1,900 fpm
Maximum speed	326 mph (100% power)
Max cruise speed	302 mph (75% power, at 19,000 feet)
Range at max cruise speed	1,900 miles
Speed for max range	215 mph (37% power, at 20,000 feet)
Maximum range	2,750 miles (37% power, at 20,000 feet)
Stall speed	102 mph

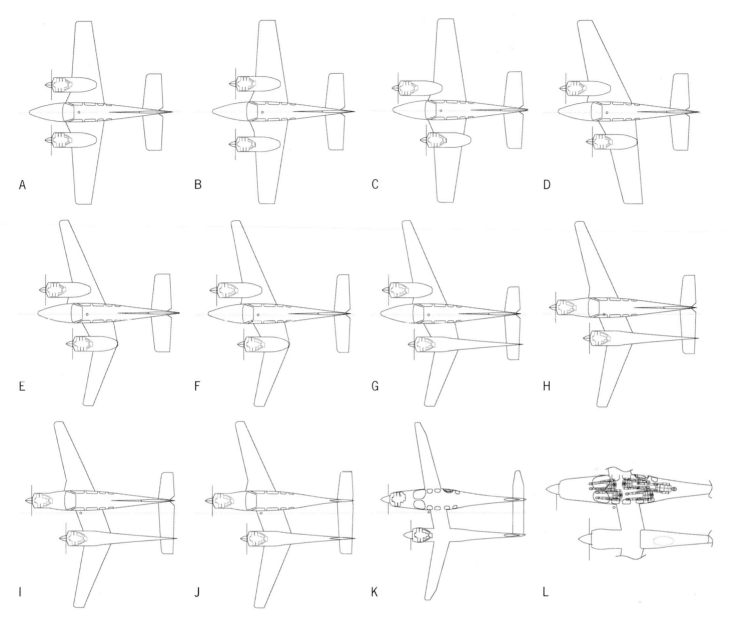

A   B   C   D

E   F   G   H

I   J   K   L

This diagram and description from Burt Rutan explain how a symmetrical twin-engine aircraft evolved into an asymmetrical twin-engine aircraft, starting the sequence with (A) the baseline of a Beechcraft Baron Model 58P. The small circle behind the windshield represents the center of gravity. (B) The left engine moved outboard to improve symmetry at low speeds and to reduce cabin noise. (C) Both engines moved inboard to reduce minimum control speed. The right engine moved forward to clear fuselage, and the left engine moved aft to balance. (D) Wing skewed to support engines and to reduce left engine interference. (E) Composite construction allows smaller, higher aspect ratio wing, but configuration is now nose-heavy, thus left wing is swept forward. This helps, but configuration is still nose-heavy. (F) The weight savings allows smaller engines, and tail area can be reduced. (G) High aspect ratio tail flutter problem is fixed with nacelle boom. This allows additional baggage room in the boom. (H) The right engine is moved to the fuselage to reduce weight, cost, and drag. Lateral balance is restored by moving entire wing to the left. Minimum control speed is now well below stall. (I) The left engine is moved outboard to reduce cabin noise and to eliminate prop interference. The entire wing is moved left to restore lateral balance. (J) Twin small vertical tails improve low-speed handing, reduce weight, and allow low-drag pressure-recover aft fuselage. (K) Continued evolution: round fuselage, increase room, laminar flow flying surfaces, higher wing loading, aspect ratio to 13.2, and full-span camber control for aileron/flap/wing optimization. (L) In the Boomerang, the pilot sits in right seat, aft section is seat or bed, and baggage area in the boom is 120 inches long (indicated by dotted oval). *Courtesy of Burt Rutan*

Scaled Composites' projects are often veiled in secrecy. The company was founded in 1982 during a time when the Rutan Aircraft Factory still worked on airplanes like Grizzly and Solitaire. It had become evident as Rutan Aircraft Factory worked on programs like the NGT that its open-door policy conflicted with the need for confidentiality. Shown here behind the curtain is the beginning of SpaceShipOne and White Knight. *Mojave Aerospace Ventures, LLC. SpaceShipOne, a Paul G. Allen Project*

Chapter 3

The Scaled Composites Years

The design for the Starship, a twin pusher propeller business aircraft with a variable sweeping canard, began in 1981 by Burt Rutan for Beechcraft while he still ran Rutan Aircraft Factory. What he would eventually build and fly, though, was an 85 percent–scale version based on this design. Rutan already had experience building subscale demonstrators for NASA, with AD-1 (15 percent scale), and Fairchild Republic, with NGT (62 percent scale).

"The reason they were built at just that scale was that we could identify an available jet engine we could afford," Burt Rutan said. "And we had to build an aircraft that was the size and weight so that it would fly well with those engines."

The other important element is that subscale versions are smaller, thus, easier and cheaper to build. To get the weight of a subscale aircraft, the scale factor is cubed. So, for example, an aircraft at 50 percent scale would weigh 12.5 percent of the full-sized version.

"We built a 62 percent NGT. It had adequate thrust, and it would fly single-engine on the Microturbo engines. By putting the pilot in the center of a side-by-side airplane, he had room."

But doing a project like NGT did create some unexpected problems for Rutan and RAF. Fairchild Republic had hoped that RAF's work would win it a giant contract with the USAF, which it did indeed do. But this work was proprietary. It was work that Fairchild Republic was having RAF do to get a leg up on the competition. Fairchild Republic wanted its bid to win. So obviously it needed RAF's work kept a secret.

It was hard to keep a project like NGT quiet when RAF had a business that was trying to bring in people to sell them homebuilt kit plans and to give them builder support. RAF welcomed homebuilders in to troubleshoot as well help teach them how to work with composites. On one hand, RAF had an open house, but on the other, it had one curtain in the corner that it didn't want the public looking behind. Besides, the resulting flight test program of confidential work could not be completed indoors.

Painting by Stan Stokes

Designs from Scaled Composites, 1983 to 1991
1—Pegasus
2—Starship (Model 115)
3—Microlight (Model 97)
4—ATTT (Model 133)
5—ATTT (Model 133-B)
6—Triumph (Model 143)
7—ARES (Model 151)
8—Scarab
9—CM-44 (Model 144)
10—Predator (Model 120)
11—Pond Racer (Model 158)
12—Stars and Stripes Wing Sail

Plans End and Plans Begin

Over the first three years of Scaled Composites' existence, Rutan felt more and more that the homebuilt market was reaching a saturation point for his designs. He could only sell so many sets of plans. And for the few hundred dollars he'd gain for each, it would cost countless hours in builder support. His business model for RAF would soon become unsustainable. And there were other concerns for RAF. Rutan had to make a very hard decision.

"We shut down the business in 1985 mainly because I was doing two businesses and this one had the large product liability risk," Rutan said.

While product liability and consumer protection are very important, frivolous lawsuits can crush a small business. If a person decides to take his alleged girlfriend on a ride, after drinking alcohol, forgets to install bolts used to reattach the wing, and is killed along with her from the resulting accident, is it right that the widow sues the designer of the homebuilt kit? Isn't this like getting scalded after trying to smuggle a hot cup of coffee, in your pants, into a movie theater then blaming the coffeehouse where the coffee was purchased because it was too hot?

"One of the least pleasant things you can do in life is to have to get acquainted with a lot of different lawyers. I didn't enjoy any of that. So I had to pick in 1985 whether

Range of Max Altitudes

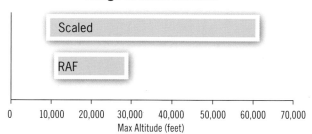

Range of Max Speeds

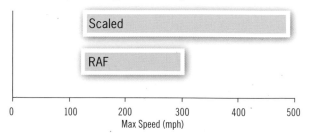

The graphs show the range of max altitudes and range of max speeds for Scaled Composites and RAF manned aircraft prior to SpaceShipOne. Compared to RAF in each graph, Scaled Composites has double the range. However, when SpaceShipOne came along, it flew at a max speed of Mach 3.25 and a max altitude of 367,500 feet—way off the charts. *Dan Linehan*

I would do the things that Scaled Composites did or I would do the things that Rutan Aircraft Factory did. And the main reason I chose Scaled Composites is I really didn't have anything on the burner that I thought was a breakthrough new homebuilt airplane that would be significantly better and interesting compared to the others that were out there. It came at a time when I didn't have any new ideas, and I decided to close up RAF."

In *Canard Pusher* No. 45, July 1985, Rutan announced that RAF would stop selling homebuilt plans. However, RAF continued with builder support for twenty or so more years.

Not only did RAF undergo a big change, but so did Scaled Composites. That same year, Beechcraft purchased Scaled Composites.

"Anyway, '85 comes along and Jim Walsh takes over as president of Beechcraft," Rutan said. "And he announces to me that he wants us to get involved in a whole variety of airplanes, not just the Starship."

Vehicles in all Shapes and Scales

"The business that I founded in '82, Scaled Composites, is very different," Burt Rutan said when comparing Rutan Aircraft Factory during a talk in 2009 at the Art Center College of Design in Pasadena, California. "We've developed, so far, twenty-eight manned airplane types. The concepts were done inhouse for twenty-four of them. Some of the concepts came to us, and we built their design. We did three company-funded research programs. Six of these types, so far, and again, this is just the manned airplanes, were for U.S. government customers. Four had a prime aerospace-based customer. Five were foreign customers. And we did no marketing at all to the public."

When the fifteen manned airplane types built under the banner of RAF and the Scaled Composites' list with one new addition are considered, over a thirty-nine year period, from the VariViggen in 1972 to SpaceShipTwo in 2010, forty-four manned aircraft types in total were built. That is an astounding statistic—greater than one new aircraft per year.

It is important to note that not every one of these vehicles was completely designed from the ground up and flown entirely by Scaled Composites each time. In some cases an existing aircraft type like the Long-EZ was modified to evaluate a new engine, as with the Jet LEZ Vantage and Pulse-Detonation LEZ. And for Roton and the balloon gondolas, Earthwinds and Global Hilton, Scaled Composites focused on their airframes.

"I did do all the designs myself at RAF, and for most of the aircraft that Scaled developed up to about ten years ago," Rutan said. "All the Scaled designs had other designers for systems and for structural details, but I was responsible for the design concepts and the basic preliminary designs."

Scaled Composites has continually brought in very talent engineers and personnel. And they have made integral contributions to the vehicles built by Scaled Composites.

"Actually, for the last aircraft that I did have design responsibility—SpaceShipOne and White Knight—I did more detailed design than I did for most of the previous Scaled projects. I did do the majority of detail design releases to shop on things like landing gear, flight controls, ECS (not avionics), rocket propulsion tank and CTN, structure and manufacturing method, etc. I designed the airfoils for the SpaceShipOne wing and tail, which was unusual, since I had others do airfoils for almost all other Scaled aircraft."

Later, we will devote entire chapters to Rutan's celebrated private spaceflight breakthroughs, first with SpaceShipOne and White Knight, and later, SpaceShipTwo and WhiteKnightTwo. For now, we will provide an overview of other manned vehicles flown by Scaled Composites, then touch on unmanned and nonflying projects, providing summaries and brief descriptions of each. Refer to the paintings by Stan Stokes to get a sense of the huge variety of vehicles Scaled Composites has worked on. Finally, the book's appendix provides a partial listing of RAF and Scaled Composites projects by model number. It is a long but incomplete list.

First Flights of Manned Aircraft by Rutan Aircraft Factory (Blue) and Scaled Composites (White)

Date	Aircraft	Date	Aircraft
1972	VariViggen	1990	Lima 1
1975	VariEze POC	1991	Pond Racer
1975	VariViggen SP	1991	Lima 2
1976	VariEze Homebuilt	1991	Earthwinds
1977	Quickie	1993	Raptor D-1
1978	Defiant	1993	Jet LEZ Vantage
1979	Long-EZ	1994	Raptor D-2
1979	AD-1	1996	Boomerang
1981	AMSOIL Biplane Racer	1996	Vantage (VisionAire)
1981	NGT	1997	V-Jet II
1982	Grizzly	1998	Global Hilton
1982	Solitaire	1998	Proteus
1983	Microlight	1999	Roton
1983	Starship	2000	Adam 309
1984	Voyager	2001	Rodie LEZ
1984	Predator	2002	TAA-1
1987	CM-44	2002	White Knight
1987	ATTT	2003	SpaceShipOne
1988	Catbird	2004	GlobalFlyer
1988	Triumph	2007	Pulse-Detonation LEZ
1988	ATTT Bronco Tail	2008	WhiteKnightTwo
1990	ARES	2010	SpaceShipTwo

Manned Vehicles: An Overview

Looking like a missile nosecone fitted with a canard, sweptback wing, and pusher engine, the Microlight (Model 97) was designed for Colin Chapman, founder of Lotus, to be introduced into the microlight weight class for the British market. It first flew with a small KFM 109 engine but was planned to be powered by a new 25-horsepower, 100–pound Lotus engine. Though design started after the Starship, it was the first vehicle ever flown by Scaled Composites.

The CM-44 (Model 144) built for California Microwave was also a milestone for Scaled Composites. Looking like a Long-EZ with a blunt rectangular-shaped nose, it was Scaled Composites' first foray into remotely piloted vehicles (RPV) or unmanned aerial vehicles (UAV). However, it could be flown with or without a pilot. It was designed for military applications and had a shape that reduced its radar signature. The CM-44 had a top speed of 195 miles per hour and an eighteen-hour flight endurance.

Model 133, the Advanced Technology Tactical Transport (ATTT), also called the Special Mission Utility Transport (SMUT), was a subscale technology demonstrator made for the Defense Advanced Projects

Even at 85 percent scale, the twin, pusher engine Starship (Model 115) dwarfs previous Rutan designs. The four-place, push-pull Defiant; the two-place, pusher Long-EZ; and the two-place, pusher VariViggen are shown behind the Starship, front to back. *Courtesy of Burt Rutan*

An agriculture aircraft, Predator (Model 120) had large hopper, with a capacity of 80 cubic feet, positioned between the pilot and the engine. Since Predator had to drop its payload while in flight, losing a large percentage of weight between takeoff and landing, it had to balance over a wide range of weights. Thinking of the payload as extra fuel or a spaceship perhaps, then Predator would have to have some similar design considerations in terms of balance as would an aircraft designed to fly long-range or be a mothership. *Courtesy of Burt Rutan*

To improve the functionality of the ATTT (Model 133), it was redesigned with a bronco tail and access out the back. Shown here on the left as the new Model 133-B, these modifications made it easier for air drops and parachutists. Triumph (Model 143), on the right, was Scaled Composites' first jet. It had two FJ-44 turbofans, each providing 1,800 pounds of thrust—a big step up from the 220 pounds of thrust per engine for the Microturbo turbojets used by the AD-1 and NGT. *Courtesy of Burt Rutan*

Painting by Stan Stokes

Research Agency (DARPA). The STOL transport was intended to fill the gap between helicopters and larger fixed-wing transports.

ARES (Agile Responsive Effect Support), Model 151, a light attack aircraft, was developed as an inhouse project by Scaled Composites. With a canard, an inner wing swept back about 50 degrees, an outer wing swept back about 15 degrees, and two vertical stabilizers, it had even more intriguing features. To prevent smoke being sucked into the engine from a gun mounted on the right side of the fuselage, it only had a single air intake that was mounted on the left side of the fuselage. This resulted in an asymmetric fuselage.

Lima 1 was a program for Toyota to evaluate the use of a Lexus car engine for general aviation. Flight tested on one side of a twin engine aircraft, the engine would eventually be used for a new airplane.

The Pond Racer (Model 158) was designed as an unlimited class air racer. The intention was to develop an aircraft that would outcompete the vintage World War II aircraft that flew in these races so the dwindling stock of warbirds could be preserved. The Pond Racer looked like a pod racer from *Star Wars*. Two giant engines on

James Linehan

Designs from Scaled Composites, 1988 to 1995
1—Earthwinds (Model 181)
2—Raptor D-2 (Model 226-B)
3—B-2 RCS (Model 175)
4—Undisclosed
5—Raptor D-1 (Model 226)
6—Eagle Eye (Model 231)
7—Freewing (Model 233)
8—Lima 2 (Model 191)
9—Searcher
10—Comet
11—DC-X
12—Kistler Zero
13—Ultralite
14—Su-25 ROAR
15—Bladerunner

Unlike the Raptor D-1 where a test pilot would straddle the aircraft just like a scene out of *Dr. Strangelove*, the larger Raptor D-2 (Model 226-B) could actually fit someone inside when configured for manned flight. The D-2 was part of NASA's Environmental Research Aircraft and Sensor Technology (ERAST) program. *NASA*

booms extended out in front of the wing with the cockpit located squarely between them, the wing, and the horizontal stabilizer.

Using a Lexus five-liter automotive engine with a turbocharger, Lima 2 (Model 191) was an aircraft developed for Toyota. Lima 2 was quiet and roomy, and its propeller turned at 1,900 revolutions per minute at cruise speed. In comparison to a Bonanza, it had 2.5 times the range and was 40 knots faster with the same horsepower.

Earthwinds (Model 181) was a twenty-four-foot-long, ten-foot-diameter pressurized gondola for a balloon designed for Larry Newman's around-the-world flight attempt. Richard Branson was an early sponsor before forming his own team. To balance the lift as it floated along in the jetstream at around thirty-five thousand feet, day and night, the gondola was attached between a helium-filled top balloon and an air-filled anchor balloon, which hung below.

The Raptor D-1 (Model 226) was designed to study the possibility of intercepting ballistic missiles at high altitudes during their boost phase. First flown in May 1993 and intended as an UAV, these types of vehicles

were still in their infancy. For the early stages of flight testing, a pilot actually mounted the Raptor, like a cowboy rode a horse, to rein it in if things didn't go right.

Based on a Long-EZ, the Jet LEZ Vantage (Model 61-B) evaluated the Williams FJ107 jet engine. A striking feature of this design was a square, flat section, as seen from below, added at the back of the aircraft. The section wasn't as wide as the strakes at the front wing roots but was longer along the fuselage, in comparison.

Developed with the idea of revitalizing the light aircraft industry by moving from propellers to jets, the Williams V-Jet II (Model 271) was designed to be affordable. The aim of this light aircraft was centered on the use two new "extremely quiet, very low-cost, lightweight" Williams FJX-2 turbofan jet engines to be mounted on its V-tail, though the prototype first flew with less powerful FJX-1 engines.

In the early 1990s, Dick Rutan got into the race to be the first around the world in a balloon. He had Scaled Composites design the Global Hilton (Model 257), an eight-foot diameter, carbon fiber composite, spherical gondola. Like Earthwinds, it would use both helium and air to maintain altitude while floating along

The Williams V-Jet II (Model 271) and the FJX-2 turbofan jet engine were developed with the idea to provide the basis of a nationwide air taxi service, similar to the idea of calling for a cab in New York City. *Jim Koepnick/Experimental Aircraft Association*

Because of its very good and well understood flying qualities, the Long-EZ was used by Scaled Composites and other companies as a platform for new technology, especially engine testing. The pod underneath the belly of Borealis (Model 61-PD) contained mechanisms for an experimental pulse detonation engine. This engine generates thrust from supersonic shockwaves that result from controlled and repetitive explosions of fuel and air. *USAF*

Given a deadline of 16 November 1996, for the first flight, Scaled Composites had only eight and a half months to build the VisionAire Vantage (Model 247). The low-cost, high-performance business jet featured forward-swept wings, a single jet engine, and two air intakes mounted on the fuselage right above the wing. At a max cruise speed of 403 miles per hour, it could carry a crew of two and four passengers 1,150 miles. *Ken Lichtenberg/Experimental Aircraft Association*

Scaled Composites built the shell of Roton, a technology demonstrator that deployed rotor blades for reentry. Rocket-powered rotor blade tips allowed Roton to hover and take off and land vertically. SpaceShipOne test pilot Brian Binnie flew Roton before joining Scaled Composites. Roton is now on static display at Legacy Park at the Mojave Air & Space Port. *Mike Massee*

in the current of the jetstream. However, the helium and air would be keep in separate cells within a single envelope that carried the two-person gondola below.

The Adam 309 (Model 309) was a push-pull twin designed for very safe handling in the event of an engine out. It had a bronco tail, which is a twin tail with the vertical stabilizers on each of the two booms connected by a horizontal stabilizer. Fabrication techniques allowed it to be developed in record breaking time for Scaled Composites, and the company reported, ". . . there are several major structural components that were produced as single-cure parts. The outboard wings, horizontal tail, elevator, rudders, and flaperons had no secondary bonds in their primary structure. This allows lighter, stronger, and safer structure due to the significant elimination of fasteners and secondary bonds."

The Rodie LEZ (Model 61-R) was based on a Long-EZ and first flown on August 2001. The aircraft is currently registered to McDonnell Douglas. The original

Long-EZ received its airworthiness in 1987, but the scope of this project remains proprietary.

Scaled Composites didn't do the aerodynamic design of Model 302, the Toyota Advanced Aircraft (TAA-1), but it did assemble and flight test the aircraft. It had a low wing, conventional tail, and fixed tricycle landing gear. The four-seater was powered by a 200–horsepower Lycoming IO-360.

Designed for and flown by Steve Fossett, with sponsorship from Richard Branson, GlobalFlyer (Model 311) made three record-breaking round the world, nonstop, nonrefueled flights between February 2005 and March 2006. A trimaran with a cruciform tail at the end of each boom, GlobalFlyer had a maximum takeoff weight of twenty-two thousand pounds; however, because of the lightweight composite design, the pilot and aircraft weighed only 3,700 pounds. So on takeoff an incredible 83 percent of the entire weight was the fuel it would carry to circumnavigate Earth.

Originally designed with telecommunications in mind, the long-range, high-altitude Proteus (Model 281) has provided a platform for a wide array of applications and experiments. This workhorse is Scaled Composites' top flyer, logging between 2,500 and 3,000 flight hours as of 2009. Proteus was also intended to be a spaceship launcher. *NASA*

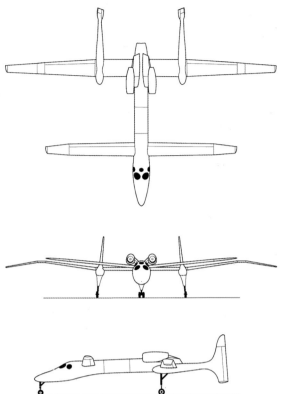

With removable wingtips installed, the tandem-wing Proteus has a wingspan of 64.7 feet and 92 feet for its front and rear wing, respectively. Powered by two FJ-44 turbofans, it has a cruise speed of Mach 0.42 and can reach an altitude of around 65,000 feet. With an empty weight of 5,900 pounds, Proteus can carry a payload weighing up to 7,260 pounds. Depending on the flight profile, it can fly up to 18 hours at a stretch. *NASA*

GlobalFlyer had a single Williams FJ-44 turbofan jet engine mounted behind the single-place cockpit. With a normal cruise speed of 288 miles per hour and normal cruise altitude of 45,000 feet, GlobalFlyer also made use of the jetstream to boost its speed as Steve Fossett circled the planet nonstop and without refueling not just once but on three separate occasions. *Dan Luft/Experimental Aircraft Association*

Designed to deliver up to a 1,000-pound payload into low earth orbit, the Orbital Sciences Pegasus made its first flight in 1989. Scaled Composites was contracted to design and build the wings, fins, and wing-body fairing. Though only a small part of the rocket, it certainly got Scaled Composites thinking about a winged vehicle that was air-launched and would fly to space. *NASA*

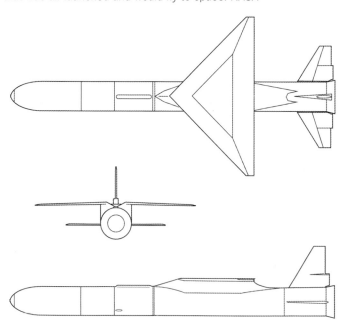

Launched horizontally at 40,000 feet, Pegasus had to "turn the corner" and point to space. For the first stage of the rocket, the wings and fins steered it. This was not too dissimilar to the flight profile of SpaceShipOne. However, the steering for Pegasus' next two stages was done by gimbaled rocket engines as it boosted to space faster than Mach 8. *NASA*

The Unmanned and Nonflying

Scaled Composites had worked early on with vehicles that could alternately be manned or unmanned, such as the CM-44 and Raptor D-1. Its work in the area of remotely piloted vehicles has steadily continued through the years.

Two fascinating RPVs, Eagle Eye and Freewing, were first flown in 1993 and 1994, respectively. The tilt-rotor demonstrator Eagle Eye built for Bell had a wingspan of more than fifteen feet, and its two wingtip-mounted rotors each had a diameter greater than nine feet. These rotors to allowed Eagle Eye to lift off vertically like a helicopter and then rotate so it could fly like a conventional aircraft. During flight testing after transitioning from hover to forward flight, Eagle Eye could hit a speed of 165 knots.

With a nose-mounted 65-horsepower engine and two opposing L-shaped tails, Freewing had a 16.2-foot wingspan and an 11.8-foot length. The leading edge of each side of the wing was connected by a shaft that ran through the fuselage. The wing was free to rotate along this shaft, with the trailing edge moving up and down with respect to the fuselage during flight. This had the effect of allowing the wing to keep a constant angle of attack whether it was climbing, maintaining altitude, or descending. Freewing's design proved stall resistant, could ride out wind gusts and turbulence very well, and could enable short takeoff and landing operation.

McDonnell Douglas and NASA worked together on the Delta Clipper Experimental (DC-X), shown here on its third landing test in 1996. The DC-X was used to evaluate a single-stage-to-orbit concept as a potential low-cost reusable launch vehicle (RLV). For this concept, a vehicle would use the same rockets to liftoff, to reach space, and to land. Scaled Composites built the shell for the DC-X. *NASA*

Most recently Scaled Composites completed work on the X-47, an unmanned combat aerial vehicle (UCAV), for its new parent company, Northrop Grumman. But Scaled Composites projects haven't been confined to the realm of aircraft. Some of the same principles behind aerospace apply to other technologies that move through the air.

In 1988, Scaled Composites made 85-foot-tall and 105-foot-tall, rigid-wing sails for the *Stars and Stripes* catamaran sailboat to race against a large New Zealand mono-hull in a special America's Cup challenge race. The GM Ultralite concept car was built in 1991 and unveiled in 1992. Made of ten pieces, the lightweight composite body of carbon fiber, sandwiched around a rigid foam, helped reduce the total structural weight down to an incredibly low 420 pounds. And in 1994, Scaled Composites designed the rotor blades for the Zond Z-40 Bladerunner, a wind turbine that had a rotor diameter of 130 feet.

The one thing all these projects had in common was the need for a high-technology solution, be it materials, manufacturing, quick turnaround, or innovation.

Painting by Stan Stokes

James Linehan

Designs from RAF and Scaled Composites, 1988 to 2000
1—Catbird (Model 81)
2—Boomerang (Model 202)
3—Nosejob
4—Adam 309 (Model 309)
5—Vantage (Model 247)
6—V-Jet II (Model 271)
7—Proteus (Model 281) with spaceship
8—Alliance (Model 287)
9—Proteus (Model 281) with antenna
10—X-38 (Model 276)
11—Global Hilton (Model 257)
12—PLADS/Rockbox (Model 179)
13—Roton
14—RCS models
15—Orion

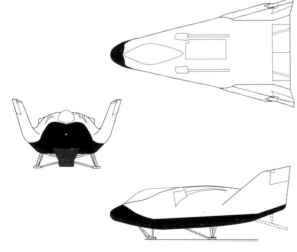

The X-38 was designed to stay attached to an airlock on the International Space Station. If an emergency arose, up to seven astronauts could board the X-38 and evacuate. A de-orbit burn would slow down the X-38, and then it would reenter similarly to the Space Shuttle, pop a parachute, and extend landing skids. *NASA*

Right: Dropping off a NASA B-52, this 80 percent-scale prototype of the X-38 (Model 276) was making it eighth—and final—freefall flight test in 2001. The X-38 was a crew return vehicle (CRV) or emergency lifeboat for the International Space Station based off the X-24 lifting body and built by Scaled Composites. The program was cancelled in 2002. *NASA*

Made of high-strength, low-weight composites, the spindly White Knight and the stubby SpaceShipOne had design configurations like no other aircraft or spacecraft. Their far-out shapes seemed like the vehicles were pulled straight from the pages of science fiction. *Tyson V. Rininger*

Chapter 4

SpaceShipOne and White Knight: The Ascension

Tom Wolf's book *The Right Stuff* chronicled the heyday of x-planes, the x-plane pilots, and the path of the Mercury Seven—the first American astronauts. In the movie adaption, while at Pancho's Happy Bottom Riding Club, the liaison officer asks Chuck Yeager (the real Chuck Yeager makes one of his cameos as Fred the bartender in this scene) and his flight engineer, Jack Ridley, "You know what really makes your rocket ships go up?"

"The aerodynamics alone are so complicated—" Ridley starts to reply.

"Funding. That's what makes your ships go up. I'll tell you something, and you guys, too," the liaison officer pulls soon-to-be Mercury astronauts Gus Grissom, Gordon Cooper, and Deke Slayton into the conversation. "No bucks, no Buck Rogers. Whoever gets the funding gets the technology. Whoever gets the technology stays on top."

These quotes relate to a substantial challenge that would have to be solved before any forward progress could be made. How can getting to space possibly be done if you are not NASA or some other giant nation with vast resources of both money and manpower?

SpaceShipOne

While SpaceShipOne in its completed form may look very beautiful to some, it certainly may not be the most elegant looking vehicle that Rutan has designed. If, however, one thinks of design elegance not as a look, but as an air, not as rich, but as refined, as something beautifully intricate yet simple, effective, and clever, SpaceShipOne fits the definition.

To start with, the engineering criteria and solutions to reach space were really governed by three fundamental principles. First, SpaceShipOne had to be cost effective. Bang for the buck may be a better way of saying it because to reach space you need a pretty big bang or a pretty big controlled explosion, as is the case for a rocket engine. The challenge was not to overdesign. Do not mistake avoiding overdesign with flying by the seat of your pants. But if you don't need all the bells and whistles to get to the goal you're trying for, then don't put in all the bells and whistles. It just costs more. It is just more things to manage, more things to maintain, and something else that can potentially fail, adding chaos into the whole system. If your car only needs 87-octane gas, then pumping high-octane gas into your car is not going to help. It's just going to cost more to operate.

Reusability

Can you imagine the cost of an airline ticket if every airliner was tossed out after each and every flight? The key to operations is to be able to efficiently reuse systems, not dispose of them. Certain irreplaceable consumables, such as propellants, are an obvious exception. The object is not to replace the whole heat shield, for example, or throw away the whole first stage of a rocket after a single use. One-shot systems don't scale well to large markets. It would not have been much of a challenge for Burt Rutan to build a single-use spacecraft.

The idea of reusability is like a car in a garage. Check the fluids and gas. Check the tire pressure. Drive back and forth. What if you had to get rid of your car after every time you took a trip? It wouldn't be economical to own a car or to even use a car. By being able to reuse as much as possible, you have systems in place where it is essentially plug in the nozzle, squeeze, fill with the gas, and you're good to go.

Safety

SpaceShipOne had to be safe. This seems like an obvious thing, but if you are going to ride this thing and you're hoping you are developing a spaceship that other people will buy into and ride, then people can't be getting killed all the time or even hurt. Granted, it is space. There are challenges. Airliners have crashes. Cars crash every day. Thousands and thousands of people die in car crashes every year. That's because cars are moving fast and there are mechanical parts and there are human operators. Things are going to happen. It is just inevitable. However, you have to have a realistic level of safety. If you scuba dive, mountain climb, or

participate in any other activity that challenges the limits of the human body in an unforgiving environment, there is always risk. That will never, ever go away.

If you were to put your key in the car's ignition and thought as you drive away from your house, *You know there is a very good chance that I'm never going to make it back*, then very few of us would be driving. If you take a rocket ride and said, *Hmmm, there is a pretty good chance I won't make it back alive*, then that would be a system that will struggle to sustain itself after a couple of accidents. So safety, even at an experimental level like for SpaceShipOne, was critical.

This does not mean that everything will be perfect and there won't be accidents and mishaps. It also does not mean that everything is a hundred percent overdesigned with safety factors like putting padding on everything that doesn't need to be padded or ridiculous signs and placards and extras well above and beyond rational safety measures. It doesn't mean that at all. It means that, fundamentally, the design is something that you feel safe in to go up and come back down. You would feel it is safe for your family to ride in. For a builder or an operator or someone with intimate knowledge of the spaceship, they have a lot more comfort with it. Those without familiarity need to feel safe and be safe, though.

Safety must never be compromised just for saving the bottom line. Once decisions about safety are made on bottom line only bases, it becomes a slippery slope where trouble salivates. It is just a matter of time before trouble takes a huge bite.

Whether Rutan consciously realized or not, or how far up in front in his mind this was, the success of SpaceShipOne would dictate the direction of how commercial space travel would go. If Burt Rutan failed, then who could do it? That's a heavy weight.

If you don't need a spacesuit because you're not going to be in space that long, don't use a space suit. Is not wearing a spacesuit compromising safety for cost effectiveness? That is exactly the interconnection between safety, cost effectiveness, and reusability that needed to be balanced.

Having a backup parachute system for the whole vehicle? Redundancies upon redundancies? After a while, complication inherently can introduce more problems into the system than it tries to solve.

Carried aloft by a NASA B-52 and then released in midair like so many of the X-planes before it, the X-15, a winged rocketplane, made two suborbital spaceflights—both in 1963—with apogees of 65.8 miles and 67.1 miles. Only two years earlier, the first two U.S. manned spaceflights, which were part of the Mercury program, had flown suborbital spaceflights as well but reached apogees of 116.5 miles and 118.3 miles. *NASA*

The Birth of a Spacecraft

"I always had the space bug, keep in mind," Rutan said. "When did I jump in and do it myself? It came in a time when I thought I could do it. And it wasn't with SpaceShipOne at first. I was going to do a capsule and launch it from an airplane that did a steep climb and a parachute recovery."

When Rutan started sketching out his early ideas in 1993, he was using ideas that were not truly novel. There wasn't much research involved in it. Rockets had been done before. Shooting a one-man rocket to space didn't seem like an advance. It was more of a step back to him than a step forward, even though the launch style would not be a traditional ground-launched rocket. Air launch had been done previously with the winged X-15 carried by a B-52 mothership. The X-15 did make suborbital flights above Kármán line, which is the internationally recognized boundary of space.

Elegance in Engineering

- Design out Problems
- Innovation
- Practical Thinking

- Feather
- Air Launch
- Hybrid Rocket Engine
- Cantilever Mounting
- Composite Construction
- Portholes for Windshield
- Hand-Flown
- Front Landing Skid
- Common Construction
- Common Fit Characteristics

- Size Reduction
- Lightweight
- Multifunctional
- Strong
- Simple
- Efficient
- Reliable
- Economical

- Cost Effective
- Reusable
- Safe

Dan Linehan

There is probably nothing more that Rutan likes than to compete and have a challenge with a target in his sights. Especially, if someone says it can't be done. After the announcement of the visionary Ansari X Prize, as aviation prizes had done in the past, it fueled, sparked, and set the bar for competition at space height.

However, it was Rutan's decades of experience in aerospace, his mind for innovation, and adaptable approach to engineering that enabled him to balance practicality with out-of-the-box thinking. Individual systems such as the revolutionary feather reentry or the seemingly insignificant nose skid exemplified this process, making SpaceShipOne the epitome of elegance in design.

Ansari X Prize

In 1996, underneath the top of the shiny St. Louis arch, Peter Diamandis announced the running of the most ambitious, exciting, adventurous, and elevating competition since the space race between the Americans verses the Russians. It certainly wasn't as large as the superpowers duking it out. However, in terms of the potential for people to actually participate in a space program, it certainly was a giant leap.

Diamandis received his aerospace engineering degree from Massachusetts Institute of Technology and went on to study space medicine while getting his MD at Harvard University. He was obsessed with space but also felt paralyzed because there was little to no chance of him ever having an opportunity to become an astronaut for NASA. There were just too many people trying for too few spots.

Space captivated his imagination, and he wanted to be able to fly to space not just once or twice in a

Above: Inspired by the effect that aviation prizes had on the growth of the aviation industry, Peter Diamandis sought to use this as a model to foster commercial space travel. He conceived of a competition that would have people building their own spaceships for a race to space with a $10 million prize waiting back on Earth. *Dan Linehan*

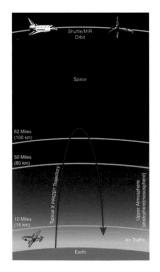

Left: In order to win the Ansari X Prize, a spacecraft had to pass the threshold of space, which was set at 62 miles above the surface of Earth. In comparison, an airliner flies about 7 miles up and the International Space Station orbits about 200 miles up. Only a suborbital spaceflight was required, but the spacecraft had to make the trip twice. *X PRIZE Foundation*

decade. He wanted to be able to fly his spaceship up to space when he wanted, the stuff science fiction is made out of.

It really wasn't that long ago when Robert Goddard, the father of American rocketry, suggested that a rocket could be built big enough to reach the Moon, and his peers ridiculed him for such folly and fanciful thought. Four decades later, Russian cosmonaut Yuri Gagarin circled Earth in orbit in his space capsule. Only eight years after that, American astronaut Neil Armstrong stepped onto the Moon.

"We needed a paradigm shift," Diamandis said. "People had become so stuck in their way of thinking that spaceflight was only for the government—only the largest corporations and governments could do this—it could never be done by an individual. This thinking was paralyzing us, and that was what I was trying to change."

After reading *The Spirit of St. Louis* by Charles Lindbergh, Diamandis found inspiration on an

Flying the *Spirit of St. Louis* from Long Island, New York, to Paris, France, Charles Lindbergh made the first nonstop transatlantic flight pursuing the $25,000 Orteig Prize in 1927. Before making this historic flight, he had little idea how far reaching his achievement would become. Lindbergh would never have imagined that three-quarters of a century later, his solo flight would provide the motivation for spaceflight. *NASA*

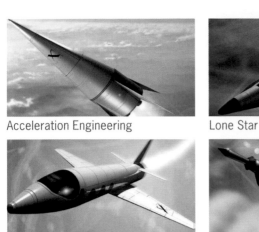

Acceleration Engineering

Lone Star Space Access

American Astronautics

Fundamental Technology

Interorbital Systems

Discraft Corporation

Pablo de León and Associates Corp.

TGV Rockets Systems

Rocketplane Limited, Inc.

The Ansari X Prize drew competitors from Argentina, Canada, England, Israel, Romania, Russia, and the United States. Twenty-six teams in total, their concepts ranged from nearly every type of spacecraft imaginable. Getting to space still proved to be quite a tough challenge, so only a few teams were able to successfully test launch hardware. *X PRIZE Foundation*

astronomic scale. The autobiographical book chronicled Lindbergh's winning of the Orteig Prize in 1927 as he became the first person to fly across the Atlantic, nonstop from New York to Paris. This was an incredible feat back then, considering aviation still was in its infancy. Airplanes crashed and pilots died as they tried to win this prize, which happened a lot during the prize flights of the early age of aviation.

No one could win the Orteig Prize at first, so it had to be reissued. Flying across the Atlantic nonstop from New York to Paris was unthinkable. Orbiting Earth was unthinkable. Landing on the Moon was unthinkable. And if the pioneers and visionaries would have just sat there and said, *Yeah, you're right. Public opinion says it's unreachable, undoable, unfathomable. Why even try?*, we wouldn't have the first Moon landing, now more than four decades ago. That was just not the kind of future that Diamandis had believed in.

There are now generations of people who have never known that at one time it was impossible to land on the Moon. After reading Lindbergh's book, Diamandis understood what it would take to change this way of thinking. And it really wasn't such a new idea. Competitions and races and prizes have been challenging people from athletics to technology to other types of advancements and, in a way, to our survival ever since the dawn of humankind. But the application and the execution of his idea was like lifting off to space itself.

People had to be infected again with the space craze. So on that day at the silvery Gateway to the West, in the city that was the namesake for Lindbergh's aircraft, the arch became a portal, an opening to the age of commercial space travel.

The Rules
The prize had to be hard enough that only serious contenders would consider it. Yet, it had to be something remotely achievable. One of the ways was to set up a system of rules that were not just a checklist but would actually help focus and encourage the type of research and spacecraft that would help Diamandis' vision come to fruition.

Had it not been for the tough rules, building a vehicle to get into space would not have necessarily been too difficult a technical challenge. If the sole purpose is just to get someone up and down safely and if someone's giving you $10 million to do it, then, with some ingenuity, it doesn't really cost that much using technology that was first developed in the 1940s.

In fact, England's Starchaser, one of the Ansari X Prize competitors, wasn't very far off of getting a single person into space. Another competitor, Canada's Red Arrow, based its spacecraft on Germany's V2 rocket, the very first human-made object to reach the edge of space way, way back in 1943.

Sending a single person into space as a one-time shot was not really much of a big deal. But if you were trying for commercialization, to make it affordable and sustainable, then you certainly couldn't afford to be sending people up one at a time on a disposable rocket.

First off, as far as the rules went, how high would a team have to go for this new space prize?

Teams had to pass the threshold of space known as the Kármán line, which is the boundary determined by physicist Theodore von Kármán, who also helped unravel the mysteries of the sound barrier. At one hundred kilometers (62.1 miles or 328,000 feet), this was also what the Fédération Aéronautique Internationale, the international organization that governs aerospace records, accepted as the start of space.

The spacecraft would have to carry three people. This was to ensure that there was room to take passengers. After all, what good was a spacecraft for commercial space travel with only a pilot and no room for passengers?

The next rule was that the spacecraft had to be flown twice in the span of two weeks. This meant the spacecraft had to be built in such a way that it was robust and durable and strong and tough enough to make the flight more than once. It couldn't be jury-rigged or duct-taped together to barely make it to space just once. The spacecraft had to go through two cycles, not just one.

Next was that only a minimal amount of the spacecraft could be replaced. Of course, the propellant had to be refueled. But for everything else, 90 percent had to be intact and could not be changed out. This forced systems that were reusable.

The spacecraft also had to be funded by nongovernment sources. If the goal was to stimulate commercial space travel, it'd be the commercial space industry that would be running the show.

And the most important rule, the crew had to be safe and sound during the launches and landings and everything in between.

The Teams

Competition was only as good as its competitors. In order for success, the field not only needed visionaries but entrepreneurs and competent talent. There also needed to be a very attractive incentive to spur their interest. Offering $10 million was quite a carrot. Back when Lindbergh won the Orteig Prize, the prize money amounted to $25,000, yet all the competitors combined spent $400,000 in pursuit of it.

The amazing leveraging of all this investment ignited the aviation industry. An aviation explosion—the Lindbergh boom—followed in talent, technology, capability, interest, and opportunity. So as important as it was to have a prize winner to ignite this new commercial space industry, it was equally important that the $10 million purse be force multiplied into an even greater economic punch.

"We probably turned away about half the applications we received," explained Diamandis "We required the teams to really demonstrate to us the seriousness of their team and effort. They had to demonstrate by virtue of the people who were involved, the companies who were involved, and they had to show us the primary concept.

"We had numerous teams apply with antigravity and UFO technology. My answer was simple: 'My office is on

Canadian Arrow

IL Aerospace Technologies

Armadillo Aerospace

ARCA

Advent Launch Services

Suborbital Corporation

To get to space, teams competing for the Ansari X Prize conceived of rockets, rocketplanes, and even a saucer-shaped spacecraft, which would ride the blastwaves of pulsejets. Some of these concepts would launch from the ground, but some would be carried to launch altitude on top or below aircraft, behind tow aircraft, or underneath giant balloons. Reentry and descent covered an equally wide range of methods. *X PRIZE Foundation*

Ever since she was a young child, Anousheh Ansari dreamed about becoming an astronaut. She had been following the progress of the X Prize, but later Ansari and her family would become the title sponsor. However, in 2006, even before her chance to ride a suborbital spaceship, she got the opportunity to fly aboard a Russian Soyuz to the International Space Station. *Prodea Systems, Inc.*

the second floor. If you can float up to the second floor, I'm happy to register you.' "

Twenty-six teams, spanning the globe, took on the challenge. Argentina, Canada, England, Israel, Romania, Russia, and the United States all fielded teams, and their vehicles and their launch methods and their return methods were as varied as in a Saturday-morning cartoon.

Some teams didn't make it much further than the paper application, but some did launch hardware, sometimes successfully and sometimes explodingly.

"The biggest challenge was, I guess, raising the finances, because the technology to do this kind of thing has been around since the 1950s, possibly even the 1940s," said Steve Bennett of Starchaser. His team had to be creative. Back in 2000, they had pre-sold two of the seats for when they would first attempt the Ansari X Prize.

Bennett said of the two prospective passengers, "They wanted to basically support the project. And they wanted to fly on the first flight. We got three seats in the capsule. The only condition they made was, 'Here's the money, Steve. We'll give you the money. We'll give it to you up front, and we're not even going to come back to you. We don't care whether it takes a year or ten years. You tell us when it's ready. We're not going to hassle you. There is only one condition.' And the one condition was the third seat had to be occupied by me. Okay. So they knew I wasn't a nutcase. They knew that I wanted to do this project and that I wanted to come home to my family."

Some even got close enough to announce that they were going to make an attempt, as with Brian Feeney's da Vinci Project out of Canada. But ultimately the funding was too difficult to secure, especially since SpaceShipOne had made one trip into space already. It certainly made it hard to bet against Rutan and Scaled Composites with their track record of safety, development of many types of flying vehicles, and work on space projects. Feeney had a successfully tested a scaled-down version of his team's launch system, which was a gigantic helium balloon. But their spacecraft, a rocket called *Wildfire*, was only 80 percent complete.

A big chunk of money came from an internet gambling company. But it wasn't the jackpot haul the da Vinci Project would need to run with Rutan and Scaled Composites.

The Prize Money

Running a highly technical prize that had merit, prestige, and integrity was not inexpensive to do by any stretch. And to raise $10 million in prize money would prove nearly as tough as it was to get into space.

At a time when the prize looked like it would not get off the pad, Anousheh Ansari stepped in. Ansari was born in Iran and always dreamed of one day becoming an astronaut. At sixteen, she immigrated to the United States then learned English. After earning her college degree, she helped found a multimillion-dollar company. After all this time, though, space was never far from her mind.

In 2001, Diamandis read a profile about Ansari in *Fortune* magazine, where she had expressed "her desire to board a civilian-carrying, suborbital shuttle."

"I read that like three times," Diamandis said. "So I convinced myself that it really said suborbital flight."

Diamandis arranged a meeting with Ansari and her bother-in-law, Amir Ansari, who was also enthusiastic about space travel.

"From the first moment we sat across the table and started to talk about it, Peter had us sold," Anousheh Ansari said.

Although she backed the prize starting in 2002, it wasn't until May 2004 that Ansari became the title sponsor. "Our sponsorship was absolutely needed for X Prize to succeed," she noted.

"At the time we joined the organization, if we had decided not to, I don't know if they would have survived. We felt that we couldn't let that happen. This was too valuable. It was difficult to put together such a good group of people again. The momentum was right. We couldn't just let it go. And at the same time, the reason we did it was because we love flying to space. And it wasn't like I want to do it just once, and we knew there were millions of people around the world that felt the same way. We wanted to do something to help build an industry, so this would become something that would be available, and you can do it again and again and again."

The prize money would ultimately be paid by a hole-in-one insurance policy. In return for very expensive monthly payments, an insurance company bet against anyone winning the Ansari X Prize.

Ansari recalls people thinking she was nuts to be supporting all this spaceship stuff. After a bit of luck or maybe it was just good karma, she would then have an opportunity to go to space even before her ride on a suborbital spaceship. She held steadfast to her dreams, and in 2006 when an opportunity opened up for her to ride a Soyuz to the *International Space Station*, she was more than ready to don a space helmet.

Adapting on the Fly

The Ansari X Prize had its rules, and the rules created constraints. For Burt Rutan this was no different than a customer having specific design requirements. And since one of the rules specified that the spacecraft must carry three people, then obviously the single-person rocket Rutan had originally envisioned would not work. For the next iteration, he designed a rocket to hold three people, which would be much greater in weight. This rocket would be too heavy for Proteus to heft. His three-person rocket was getting closer and closer to the status quo.

Either doing a ground launch or using a much larger lifting vehicle were Rutan's options. SpaceShipOne wouldn't be funded by the bankroll of a nation. But on the flipside, the performance requirements and its objectives were certainly much less than that required by the Space Shuttle or the Apollo, Gemini, and Mercury missions. Even the early suborbital Mercury missions flew nearly twice as high as SpaceShipOne would.

This was where cost effectiveness came in. One of the first things that really helped begin to shape the design of SpaceShipOne was if it ground launched, it was going to be much more expensive in terms of having to build a spacecraft that was bigger in order to carry enough fuel to bring it all the way from the ground to space.

Additional safety precautions would be needed in case SpaceShipOne had an accident or a catastrophic failure on the launch pad. A complicated and expensive system to get the crew out of harm's way would have to be used because there would be only fractions of a second to react if such a problem arose. Whereas with an air launch, if the problem sprung up, then the spaceship would be at launch altitude. And altitude equals time. During this period, the pilot could troubleshoot and decide whether to bail out or ride it out.

Rutan never really considered ground launch a viable option. It was clear then that a bigger launch aircraft was needed.

Rutan now had a large rocket trying to launch off of an aircraft horizontally. It takes an awful lot of energy for a rocket to make a turn from horizontal to vertical.

The idea of doing a zoom maneuver with Proteus had been considered for the single-person rocket, where the lifting aircraft would point down to pick up speed and then pitch up, angling the rocket upward but no way near close to vertical. Other ideas included the use of the drogue chute, a parachute to orient the rocket upward after being released where it could fire off vertically.

But when he came down to it, Rutan had the most experience with airplanes—vehicles with wings. A winged spaceship would even provide a little extra lift for the mothership, compared to if the mothership was toting a rocket of the same weight.

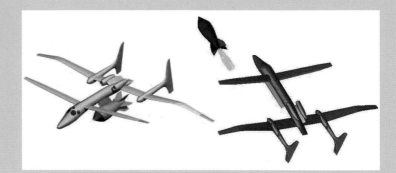

The original launch concept, from 1995, showed how Burt Rutan planned to fire a one-person rocket from Proteus to space. Proteus would perform a zoom maneuver, swooping down and then pitching its nose up, to aim the spaceship upward. The spaceship would detach, fire its rocket engine, and head to space. *Mojave Aerospace Ventures, LLC. SpaceShipOne, a Paul G. Allen Project*

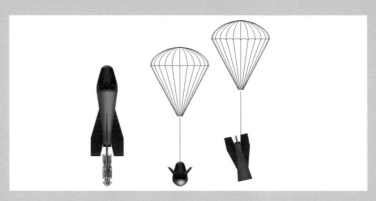

With the announcement of the Ansari X Prize and its rule requiring a crew of three, Burt Rutan created a concept for a spacecraft that could carry three people. This design was still a rocket, though. The capsule would return with a parachute, but instead of a water landing like NASA's manned capsules, Rutan intended for it to be snagged in midair. The protuberances on the capsule were designed to slow down the capsule during reentry. *Mojave Aerospace Ventures, LLC. SpaceShipOne, a Paul G. Allen Project*

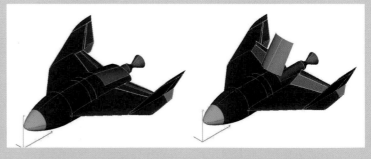

Based on all his experience with winged vehicles, Rutan then moved from rocket to rocketplane concepts. This model used a large speed brake and two large elevons for reentry. However, this design was abandoned because it could not fly both subsonically and supersonically without a complicated and expensive flight control system. *Mojave Aerospace Ventures, LLC / SpaceShipOne: A Paul G. Allen Project*

Above: After the Apollo program completed in 1972, the Space Shuttle, entering service in 1981, took over manned spaceflight for NASA. A fleet of six Space Shuttles were constructed: *Enterprise* was used for flight testing; *Challenger* and *Columbia* were lost during accidents; and *Atlantis*, *Discovery*, and *Endeavor* will all have been retired by 2011. *Dan Linehan*

Right: The Space Shuttle ground-launched and used air resistance as it entered the atmosphere from space to decelerate. A ceramic tile heat shield protected the Space Shuttle from the immense heat buildup as it hit the atmosphere at Mach 25. After reentry, it descended as a glider and then landed on the runway. This photograph shows an air-launch of *Enterprise* during the Space Shuttle's flight test program. Former NASA and Scaled Composites test pilot, Fitz Fulton, flew the 747 for some of these test flights. *NASA*

Landing Method

After reentry, there are three main methods to get back on the ground: parachutes, powered descent and landing, and gliding. Parachutes were used by all of the Mercury, Gemini, and Apollo programs. Powered descent and landing has been relegated more to R&D at this point. But there is a lot of interest and activity by companies exploring this technology in order to use it effectively and efficiently. Gliding has been used successfully by the Space Shuttle and, even before it, with the X-15.

When you think of the systems involved in each of these three types, there are some big drawbacks to some of these approaches that end up being quite a big expense.

A parachute does seem like the simplest way. A spaceship basically pops the shoot and floats down to the surface. The problem with a parachute is you don't have a controllable landing. You don't know exactly where you are going to land.

For all the rocket-based space programs of the United States, they all parachuted into the water. In this type of landing, the water helps absorb some of the shock. But the spaceship is out in the middle of the ocean. You have to contend with the recovery aspects and with all the manpower required to make sure you get your astronauts and spaceship back safely. Once it's reeled in from the water, now you have all the expense incurred to get the spaceship back home. On top of that, you have the corrosive aspect of the seawater, possibly destroying sensitive components. So it makes reusability a very big challenge.

In early design concepts by Rutan, he did consider doing a parachute return because the vehicle was quite simple and small. He felt that it would be a possibility just to snag the vehicle out of the air by an airplane or helicopter.

The White Knight mothership lifted SpaceShipOne to a launch altitude of 47,000 feet, clear above 85 percent of the atmosphere. This added a big margin of safety, compared to ground-launching, in the event of a malfunction during the ignition of the rocket engine. The spacepilot in SpaceShipOne had altitude, which equals time, to troubleshoot or decide to bailout. *Jim Koepnick/Experimental Aircraft Association*

This was very similar to a way early spy satellites used to operate. They didn't have the telemetry to communicate back and forth the way we have now. So when a spy satellite went up to space, it shot film. No one could see the images until the film came back to Earth and was developed. These satellites, after the shooting the film, would eject a film canister. It would fall to Earth, pop a parachute, and an airplane would come by and catch it in midair. But film canisters were relatively small objects. They weren't anywhere near the size or the weight of a one-, two-, or three-person space capsule.

Using a spacecraft that has its own power to maneuver after it returns into the atmosphere does solve the problem of being able to maneuver back to where you want the vehicle to end up. So you are obviously avoiding a lot of the cost in transportation and resources for recovery. However, it is an enormously wasteful process because you are spending a lot of energy to actually lift propellant to space that won't be used until after reenty. And that is hugely expensive to do.

It's not just the mass of the extra propellant needed after reentry. Now you need additional propellant to

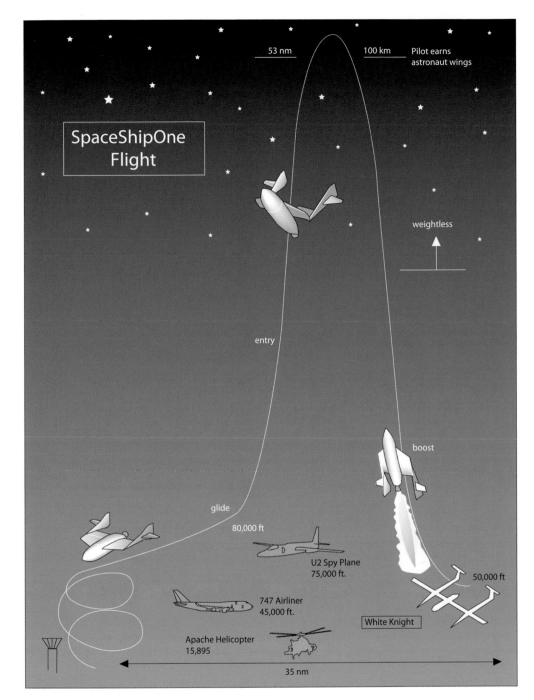

SpaceShipOne
Flight

53 nm   100 km   Pilot earns
astronaut wings

weightless

entry

boost

glide

80,000 ft

U2 Spy Plane
75,000 ft.

50,000 ft

747 Airliner
45,000 ft.

White Knight

Apache Helicopter
15,895

35 nm

compensate for this additional weight. With all that additional propellant, there needs to be more room for it in the spaceship. The additional room requires the spaceship to be bigger. So all your structures and systems have to accommodate this extra weight that now themselves require additional propellant, not just to lift but also to slow down. So this compounding mass makes it a very expensive method for returning from space. There is just no way around it.

The optimum way to return to Earth is to let gravity do the work. Yes, the parachute does do that. But if you could let gravity do the work and get you back to the point where you want at the same time, then that really

SpaceShipOne's suborbital spaceflight could be broken down into ten different phases.
1. Liftoff of SpaceShipOne mated to White Knight
2. Captive-carry to launch altitude
3. SpaceShipOne separation from White Knight
4. Supersonic boost to space
5. Coast to apogee
6. Freefall from apogee
7. Supersonic reentry into the atmosphere
8. Descent with feather still up
9. Gliding descent back to runway
10. Horizontal landing
Mojave Aerospace Ventures, LLC. SpaceShipOne, a Paul G. Allen Project

solves a bunch of problems. It becomes a heck of a lot cheaper, and that is exactly what the Space Shuttle did. It glided back home. There were no extra engines or extra propellant. It was just a matter of coming down. And that was the system that SpaceShipOne used—gliding. If a vehicle is designed such that it has a substantial glide range, then it can also accommodate a return after poor trajectory or being far off course.

A spaceship that glides home can be built smaller, which is cheaper to construct and cheaper to operate. The need for an expansive and expensive recovery procedure is also eliminated.

The Feather Concept

The *ah-hah* moment or the key to the development of SpaceShipOne boiled down to one single discovery—the feather system used for reentry. This put everything in motion for Burt Rutan. He was now ready to execute his ideas, dreams, and designs for space. Rutan had known that he could not move forward with the project until the problem of reentry was solved.

Up until this point, the modeling work that had been going on with initial concepts of SpaceShipOne had not been yielding very good results. One such model had big speed brakes to slow it down. This design had the ability to work subsonically or supersonically, but not both. SpaceShipOne would have to operate both below the speed of sound and well above the speed of sound. Rutan then developed the idea of the feather, where the whole back half of the

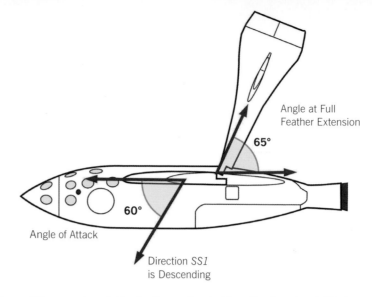

Above: When deployed, the feather extended to a fixed angle of 65 degrees prior to reentry into Earth's atmosphere. As SpaceShipOne descended, it came down nearly flat on its belly. However, it didn't fall straight down but came down moving forward at an angle of attack of 60 degrees. *James Linehan*

Opposite: An air-actuated piston on each side of the fuselage raised and lowered the feather, as shown. Having this pair was a redundancy because either one could lift and retract the feather by itself. The same was true with the L-shaped feather lock, one of which is visible in this photograph. A piston on either side could unlock and lock the feather. *Mojave Aerospace Ventures, LLC. SpaceShipOne, a Paul G. Allen Project*

Below: Both the right side and the left side tail booms rose during the feather maneuver, but they didn't act independently. As a way to reduce complexity and improve structural integrity, the feather was constructed as a unified piece. *2004 Mojave Aerospace Ventures, LLC. SpaceShipOne, a Paul G. Allen Project*

Construction of the Feather

The feather was SpaceShipOne's ticket home. If it failed to deploy, SpaceShipOne would see significantly higher loads and heating on reentry, possibly damaging its structure and overheating its leading edges. Its function required very special thought, but so did how it was built and operated.

It is unknown what the condition of SpaceShipOne would have been if it tried reentry into the atmosphere without deploying the feather. It would not be good no matter what. But could SpaceShipOne actually survive without deploying the feather? Would its composite structure, which was held together by epoxy, just melt from the heat generated during reentry?

"No, it would not melt, just damage the first ply or two of structure, requiring repair," Rutan said. "We do not know for sure if it could survive a feather-down entry. It would require testing that would be a lot more extensive than our test program. For example, it likely would be strong enough, but it might flutter, which would be catastrophic. The feather allows a flight envelope that has very low loads, thus a far less risky test/validation program. We opened up and cleared the operating envelope for feathered entry only. It is true that a feather-down entry would *reach past max Q*, but max Q is what the testing qualified. The real max Q *might* be much higher."

The feather system relied on unified construction. The outboard tail booms were connected together to form a single piece. It was impossible for one side of the tail to go up without the other. SpaceShipOne could never end up in a position where one side moves and the other doesn't move in exactly the same way. The back halves of the wings were built around the tail boom structure, so they were tied together as well. This construction eliminated a huge failure mode altogether.

The locking mechanism used the same type of idea. It was a single unified lock structure, but engaged at either side of the fuselage. So SpaceShipOne couldn't get into the position of having one lock disengage while the other lock didn't. This eliminated that failure source as well.

The locking mechanism had two independent actuating systems, on the left side and on the right side of the fuselage. Each of these independent systems was capable of actuating both locks at the same time. In the event the actuator failed, SpaceShipOne had redundancy. So it would have to have both independent systems fail not to be able to unlock and lock the feather.

The same setup was used for the actuators that raised and lowered the feather. Again, there was an independent left and right system, and each was capable of lifting the entire feather tail section up and lowering it down without the other. This was a very important redundancy. These actuators were the same type of part used by White Knight for its landing gear. Defects or systematic problems could be detected before SpaceShipOne ever flew. This gave a very good feeling of confidence and reliability, which didn't hurt for such a vital part of SpaceShipOne.

The actuators, which were pneumatic pistons used for both the locking mechanism and the feather mechanism, were powered by compressed air. The source of the compressed air for SpaceShipOne came from six pressure bottles. Each side's actuators fed off a separate pressure bottle. And each of these pressure bottles could be crossed over to supply the other side. Each pressure bottle could also feed both sides at the same time. So if there was a pressure bottle failure, the pilot had several backups. The compressed air was also used to actuate as many common systems as possible.

wings, including the outboard tail sections, folded nearly straight up like the shape of an "L."

The feather system achieved tremendous benefits in one simple design. This was a truly elegant solution. First was the fact that SpaceShipOne decelerated very quickly in the high atmosphere because of the drag. SpaceShipOne would not encounter significant heat buildup. The feather forced SpaceShipOne to fall belly first. It was like using a giant parachute because of the cross-section. Instead of the aerodynamic shape of the pointy end going head-on into the airstream, SpaceShipOne had its belly and half its wings—the parts with the most cross-section—in the airstream. This was a huge difference in drag compared to the head-on orientation.

The second benefit was that the feather operated completely hands free. The pilot only had to use one lever to lock and unlock the feather and another lever to raise and lower it. There was no steering required. There was no fine tuning of the feather's angle. The pilot activated it before reentry, deactivated it before gliding, and that was it.

Another benefit was the feather could self-right SpaceShipOne. So the spaceship could be coming from space upside down, but as it settled into the upper atmosphere and the air pressure against it started creating drag, SpaceShipOne would reorient belly first and steadied itself in that position.

This happens very much like the way a badminton birdie moves through the air after it is struck by a racket. The knobby head quickly turns into the direction of flight as it rapidly slows down.

SpaceShipOne was so stable in the feathered configuration that the pilot could hardly pitch the nose up or down. However, he could easily rotate on its belly, moving the nose right and left.

Thermal Protection System

Because SpaceShipOne decelerated so quickly and so high up in the atmosphere, there wasn't much time for it to heat up. The air was relatively thin here, so heat had little time to conduct from the air molecules outside to the skin of SpaceShipOne. As a matter of fact, Rutan had stated that the thermal protection measures were needed more on the boost stage when SpaceShipOne was going above Mach 3 in a denser air compared to the reentry.

SpaceShipOne's thermal protection system consisted of a combination of high-temperature resistant resin added to the composite at critical locations and a painted-on ablative coating. Showing up as red paint, this ablative coating was made of plastic that steals away the heat generated outside the spaceship by forming chemical reactions as it burns off.

Funding

A group of businessmen from St. Louis, Missouri, funded Charles Lindbergh for his solo flight across the Atlantic Ocean to claim the Orteig Prize.

They said to him, in essence, *Look, Slim, we'll make sure you've got the money you need. All you have to worry about is designing, building, and flying an airplane to make the flight. We trust you. We won't interfere.*

Not only did Lindbergh very much appreciate this type of hands-off support, it also turned out vital to his success. He then could focus on what he needed to do, which was hard enough already. Funding turned out to be gigantic distractions for some of the other competitors who were considered the favorites and frontrunners. They, otherwise, would have been able to fly even before Lindbergh made his attempt.

Shown after the first spaceflight of SpaceShipOne, investor and philanthropist Paul Allen (left) and aircraft designer Burt Rutan (right) had formed a partnership called Mojave Aerospace Ventures in 2001 to develop a privately funded space program.
Tyson V. Rininger

In 2001, construction began on SpaceShipOne (Model 316) and White Knight (Model 318) under the secret program name of Tier One. Burt Rutan had categorized projects as Tier One, Tier Two, and Tier Three in terms of their fun factor, with Tier One being the most fun. He wanted to make a statement that the SpaceShipOne program would be the most fun of all, so he called it Tier One. *Mojave Aerospace Ventures, LLC. SpaceShipOne, a Paul G. Allen Project*

For Paul Allen, he'd wondered if he would ever have the opportunity to take part in a space-related initiative. Allen was also such a science fiction fan that he built the Science Fiction Museum in Seattle, Washington. But as the cofounder of Microsoft, he was also pretty serious about technology. As a kid, he read sci-fi, built model rockets, and watched Mercury, Gemini, and Apollo launches on television at school. "When the SpaceShipOne opportunity came up, I was very excited to pursue it."

Burt Rutan had been brainstorming about a spaceship since 1993, but he had a big problem to solve before seeking funding. He had to figure out how to get his spaceship back from space safely through reentry.

"My first couple of meetings with Paul were not about space at all," recalled Rutan. "There was an interest that he had in something else I was doing. It was related to Proteus for telecommunications." Rutan had been unaware of how much of a big space and science fiction fan Allen was.

When Rutan discovered that the feather would work, he approached Allen. In 2000, he told Allen that he could indeed make a spacecraft. Paul Allen immediately stuck out his hand to shake on the deal. Allen's company Vulcan (a clue to Allen's sci-fi roots?) and Scaled Composites finalized a partnership called Mojave Aerospace Ventures in 2001.

Allen put up approximately $25 million.

"There were two ways for me to recoup my investment," Allen said. "One was the winning of the X Prize, and one was the licensing we'd be able to achieve with a company like Virgin Galactic. Those were the possible future mechanisms of payment back when we were evaluating all this stuff. You didn't necessarily assume you were going to win. And you didn't know what the other competition was like."

When Doug Shane, the new president of Scaled Composites, looked back at all the key factors to the success of SpaceShipOne, he said, "It starts with a customer. We didn't have a customer in this case. We had a sponsor. They wanted us to accomplish a goal. And they didn't care how we did it. They trusted us to figure out how to do it and learn along the way. And that was amazing."

Tier One had six main components: the suborbital spacecraft SpaceShipOne, the spacecraft launch system White Knight, the test stand trailer (TST) used to develop the rocket engine, the mobile nitrous oxide delivery system (MONODS) used to store nitrous oxide and fill the oxidizer tank, the portable ground control Scaled Composites unit mobile (SCUM) truck, and the flight simulator developed specifically for the program, which is not shown here. *Mojave Aerospace Ventures, LLC. SpaceShipOne, a Paul G. Allen Project*

Rocket Engine

Burt Rutan did not have much experience with rocket engines. He spent a significant amount of time researching the best engine to use for SpaceShipOne. Two types of rocket engines were most often used in spacecraft: rocket engines with solid-based propellant and those with liquid-based propellant.

The propellant of all combustion-based rocket engines must contain a fuel and an oxidizer. The chemical reaction that creates the thrust for a rocket engine is just like the chemical reaction when wood (the fuel) burns in air (the oxidizer). The combustion reaction expels hot gas out the end and causes the reactionary movement of the spacecraft. So exhaust gas goes out one direction and the rest of the spaceship goes in the opposite direction.

One of these rocket types was very easy to eliminate as a possibility. Solid rockets have the fuel and the oxidizer premixed. Light the fuse, and the rocket engine burns. It does not stop burning until all propellant has been consumed. This, by far, is the simplest type and the cheapest type of rocket engine. The propellant mixture makes it a very efficient rocket engine as well.

This all satisfied the cost-effective element and addressed reusability because only the propellant would be replaced, as with the solid rocket boosters (SRB) from the Space Shuttle. The SRBs parachuted back after a launch and were refurbished and refueled.

However, safety turns out to be the biggest disadvantage in the case of the solid rocket engine. Note that the SRBs were not even directly attached to the Space Shuttle and could be easily jettisoned.

Imagine if you were in a flight test program, you lit a solid rocket engine off, and something did not go as planned. The vehicle could be in a very dangerous situation. For example, if the test pilot loses control, the rocket engine would continue to accelerate the vehicle, making a bad situation much, much worse. The problem is that solid rocket engines cannot be shut down before they burn out. Once they are lit, they keep on burning. So this factor immediately eliminated this type of rocket engine from the selection process.

The obvious choice would then be to use a liquid rocket engine, which was the same type used by the main engines of the Space Shuttle.

It is a little funny that they are called liquid rocket engines because many of the different liquids that act as the fuel and the oxidizer do not exist as liquids under the normal conditions that we humans walk around in, as far as temperature and pressure go.

They are heavily pressurized and chilled. So, for example, hydrogen and oxygen, which we typically think of as gases, were the propellant used by the Space Shuttle's main engines. Yet, they are liquid rocket engines. This is because liquid hydrogen and liquid

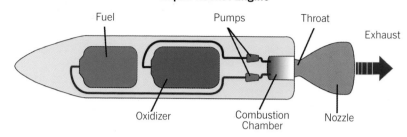

Liquid Rocket Engine

Fuel · Pumps · Throat · Exhaust · Oxidizer · Combustion Chamber · Nozzle

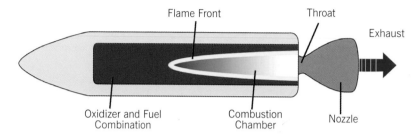

Solid Rocket Engine

Flame Front · Throat · Exhaust · Oxidizer and Fuel Combination · Combustion Chamber · Nozzle

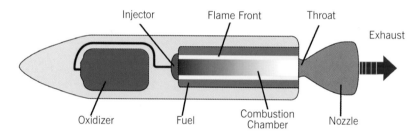

Hybrid Rocket Engine

Injector · Flame Front · Throat · Exhaust · Oxidizer · Fuel · Combustion Chamber · Nozzle

The three main rocket engine types are liquid, solid, and hybrid. Each requires a fuel and an oxidizer. The states of these propellants are what distinguish the rocket engine types. Both liquid and hybrid rocket engines have their fuel and oxidizer separated prior to ignition. However, a solid rocket engine has its fuel and oxidizer premixed. *James Linehan*

This cutaway of the SpaceShipOne fuselage shows the position of the cockpit on the left, the oxidizer tank in the center, and the case/throat/nozzle (CTN) assembly mounted on the right side oxidizer tank. Called cantilever mounting, the connection between the oxidizer tank and the CTN was the only place the rocket engine attached to SpaceShipOne. *Mojave Aerospace Ventures, LLC. SpaceShipOne, a Paul G. Allen Project*

The Space Shuttle launches into space using two different types of rocket engines. It has three shuttle main engines (SME), which use liquid oxygen oxidizer and liquid hydrogen fuel, fed from the large external tank during boost. It also has two solid rocket boosters (SRB) that are jettisoned once they burn out. The SRBs are refurbished, but the external tank is not reused after being jettisoned. These powerful rocket engines got the Space Shuttle to an altitude of around 200 miles in 8.5 minutes. *Dan Linehan*

A larger rocket engine nozzle (25:1 expansion ratio) was used by SpaceShipOne for high altitude test flights compared to ground tests (10:1). But both were ablative nozzles designed to slowly erode to help keep them from completely melting or burning up due to the intense plume of fiery gas coming out from the rocket engine. *Mojave Aerospace Ventures, LLC. SpaceShipOne, a Paul G. Allen Project*

oxygen have been cooled and kept under pressure, so they have changed their state of matter from gas to liquid, very similar to the way steam condenses to form liquid water.

The temperatures of liquid hydrogen and liquid oxygen without any pressurization are -253 degrees C (-423 degrees F) and -183 degrees C (-297 degrees F), respectively. To handle such extremely cold propellant takes very special pumps and very special plumbing and very special rocket engines.

This all adds up to a huge amount of cost and a very complicated system because not only are systems in place for normal operation, there must be systems in place as backups. Safety systems also must be in place. Although liquid rocket engines are very expensive and very complicated, liquid engines are very efficient in terms of the amount propellant by weight compared to the amount of thrust produced by burning them.

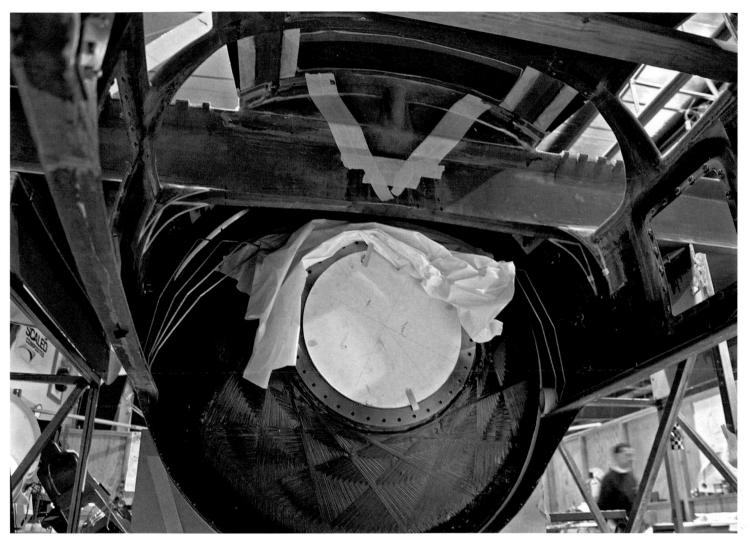

Hybrid Rocket Engine

Burt Rutan continued his search for a suitable rocket engine. At one point, a manufacturer wanted to demonstrate the function of a cryogenic valve used for its rocket engine. Liquid nitrogen, at -196 degrees C (-321 degrees F), chilled the valve to operating temperature.

"The valve stuck," Rutan said. "It didn't work. It failed right there while I was looking at it."

This did not instill confidence. He asked, "Gentlemen, let me ask you, how many of those valves are in this motor?"

Twelve, they replied.

"Did you expect this to work today?" he asked.

The answer was silence.

Before Lindbergh soloed the Atlantic in a single-engine aircraft, many of his critics charged that he should use a multiengine airplane. Lindbergh felt that

Liquid nitrous oxide stored at cryogenic temperatures was loaded into the oxidizer tank. By allowing it to warm up to room temperature, the oxidizer tank self-pressurized. The oxidizer tank had an inner fiberglass liner and was wrapped in high-strength, lightweight carbon fiber filament. It was sized to perfectly fit the inner diameter of the fuselage and acted as an important structural member of SpaceShipOne. *Mojave Aerospace Ventures, LLC. SpaceShipOne, a Paul G. Allen Project*

having two or three engines made it two or three times more likely to have an engine failure during the flight.

So the valve Rutan just witnessed malfunction was part of a rocket engine that contained twelve of these valves. That gave him worries. If he could find a rocket engine with only one valve that worked at room temperature, then that's the one he wanted.

Rutan sought to find a rocket engine that wasn't so complex and expensive. But also it couldn't be heavy because backups and redundancies add weight very fast. And every time the weight increases, more propellant is needed and then a bigger vehicle has to be built, then more propellant and so on. Weight gain could quickly go out of control, almost like an endless loop.

A third type of rocket engine that had never been used in manned spaceflight before began to get Rutan's attention—a hybrid rocket engine. This type of rocket engine is basically part liquid rocket engine and part solid rocket engine. The more a hybrid rocket engine was explored, the more things made sense and fell into place. For SpaceShipOne, the performance requirements were on the low side. So the need to use complex and expensive chemicals that produced a great deal of efficiency was not necessary. SpaceShipOne could get away with having a less efficient or a less effective rocket engine.

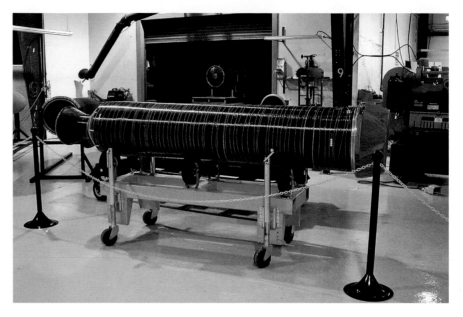

After a spaceflight, the case/throat/nozzle (CTN) would be replaced with a freshly fueled one. Although it was filled with solid rubber, it did have ports that ran the length of the tube so that when the rocket engine was ignited, the combustion gases could escape out the back through the nozzle. The flange on the right, shown partly covered by a protective cap, was used to bolt the CTN onto the oxidizer tank. The wire shown circling the CTN was a burn-through indicator. *Mojave Aerospace Ventures, LLC. SpaceShipOne, a Paul G. Allen Project*

Scaled Composites held a competition between SpaceDev and eAc to design parts of the rocket engine. Valving at the front of the oxidizer tank was designed by eAc, and the fuel and fueling of the case/throat/nozzle assemblies were provided by SpaceDev. The hybrid rocket engine used in SpaceShipOne had an oxidizer of nitrous oxide (N_2O) and fuel of synthetic rubber (HTPB). It produced 16,800 pounds of thrust. *Mojave Aerospace Ventures, LLC. SpaceShipOne, a Paul G. Allen Project*

"Would I use a hybrid motor to go to orbit? Probably not unless we could develop one that was close to the efficiency of the liquids," Rutan said.

So how do you really keep it simple as dictated by the engineering KISS (keep it simple stupid) principle?

For starters, Scaled Composites chose hydroxyl-terminated polybutadiene (HTPB) for the fuel and liquid nitrous oxide (N_2O) for the oxidizer. These two chemicals are better known as tire rubber and laughing gas. This choice sacrificed performance for simplicity, which was okay.

For a hybrid rocket engine, the fuel still needs to come in contact with the oxidizer; otherwise, there will be no combustion. The oxidizer needs to be pumped across the solid fuel. So, again, you are talking about pumps and valving and everything else it takes to move the oxidizer from the tank to the fuel.

Potentially there are a lot of systems, and each system would add not just to weight but a possible source of a problem. For example, a leak in plumbing could be catastrophic. So you have to double contain tubing and you need redundancies. If there was a leak, then how would you compensate for it? Or detect it? You need sensors around the plumbing.

To solve the problems of (1) the complexity of extra systems, (2) the weight of extra systems, (3) the cost of all the extra systems, and (4) the inherent risk to safety with extra systems, you simply reduce the number of the systems. And that was just what Rutan did.

Hardly a better illustration of elegant design is there than this rocket engine. Rutan came up with a very ingenious idea for its construction using a cantilever design. What the design called for was to attach a tube of rubber fuel directly to a tank containing the liquid oxidizer.

Any chemical—including water that freezes at 0 degrees C (32 degrees F) and boils at 100 degrees C (212 degrees F)—can be dangerous to handle under certain conditions. Steam boilers used to heat buildings have exploded before. But dealing steam (gaseous water) is a lot easier than dealing with highly toxic or highly reactive chemicals.

When liquid nitrous oxide is unpressurized, it has a temperature of -88.5 degrees C (-127 degrees F), which is much warmer than liquid oxygen and hydrogen but

The SpaceShipOne pilot ignited the rocket engine about 10 seconds after dropping off White Knight. Less than 10 seconds later, SpaceShipOne was already moving faster than the speed of sound. The rocket engine was shut off after about a minute and a half—about halfway up apogee—and SpaceShipOne coasted the rest of the way up. *Ron Dantowitz, Clay Center Observatory, Dexter and Southfield Schools*

far colder than a room temperature of 22 degrees C (72 degrees F). But when heated in a sealed vessel, such as the oxidizer tank, it self-pressurizes. Because the oxidizer tank is now pressurized, it can squirt the nitrous oxide from the tank to the fuel without the need of a pump.

Nitrous oxide is relatively inert at low temperatures, and for it to react with the rubber fuel, an igniter inside the rocket engine had to heat it above 300 degrees C (570 degrees F).

The one valve that controlled the flow of the oxidizer into the fuel—on and off—was located inside of the oxidizer tank itself. This approach allowed two significant things. It eliminated the potential leak paths because the valve was now located in the actual tank itself and it also helped control the temperature of the valve since it was bathed in nitrous oxide. Rutan reduced a whole bunch of plumbing and all the associated complexity, weight, cost, and risk.

There are other additional savings in this rocket engine design. Since the fuel tube was mounted directly to the oxidizer tank, there were no other additional supporting structures needed. Again, this extra stuff didn't have to go into the spaceship. The oxidizer tank was pressurized and had to be very strong. Rutan designed the oxidizer tank to fit perfectly inside the fuselage. So by bonding the oxidizer tank to the fuselage, it acted just like a structural member supporting the fuselage. Again, a whole bunch of extra material and potential failure points were eliminated. This was all cost reduction due to simple solutions.

Flight Controls

SpaceShipOne flew in three flight regimes: (1) subsonic in the atmosphere, (2) supersonic in the atmosphere, and (3) spaceflight free of the atmosphere. To safely accomplish this, SpaceShipOne had to employ four unique flight control systems.

Once SpaceShipOne was released from the carrier aircraft at about forty-seven thousand feet, it glided before firing off its rocket engine. So, like an airplane, it must be able to be controlled in flight. SpaceShipOne did so by using a mechanical system similar to what is found in a small aircraft, such as a Cessna 172—a single-engine, propeller-driven, four-person aircraft. The pilot had rudder pedals and a control stick that allowed him to change the pitch, yaw, and roll of SpaceShipOne just as any normal airplane does.

When SpaceShipOne's rocket engine was ignited, it quickly built up speed but initially flew below the speed of sound, subsonically. As it continued to increase speed under rocket power, the pilot still used the mechanical flight control system. These flight controls had no augmentation. They were simply mechanical rods and linkages with no powered assistance to help the pilot. The rapid acceleration moved SpaceShipOne faster and faster. Soon the force from air resistance pushing against the spacecraft was just too much for the pilot to physically maneuver against.

Imagine when you put your hand outside the window when you are traveling six miles per hour in a car. You feel a light breeze. When you put it out at

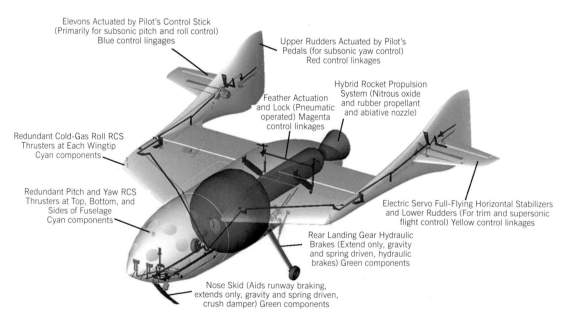

Because SpaceShipOne flew in different flight regimes, it required different flight control systems. SpaceShipOne used a mechanical flight control system for subsonic flight, an electric motor-driven flight control system for supersonic flight, a reaction control system for spaceflight, and the feather for reentry. *Mojave Aerospace Ventures, LLC. SpaceShipOne, a Paul G. Allen Project*

Hand Flying

SpaceShipOne was hand-flown. It did not rely on a fly-by-wire system, which is basically a computer, sensor, and actuator network that assists pilots flying aircraft, such as fighters and airliners, nowadays.

A fly-by-wire system could have potentially rolled the four flight control systems SpaceShipOne used into one simple to operate flight control system. However, there would have been enormous costs associated with purchasing and developing the equipment, not to mention the huge weight gain for all the computers, instrumentation, and automatic mechanisms.

All that extra weight kept SpaceShipOne that much closer to the ground. The other big drawback with a fly-by-wire system is if it goes out, the pilot has little to no chance of controlling the vehicle without it. So there needs to be a full backup of the systems as well. Again, this translates into big cost.

Chuck Yeager was the first to break the sound barrier back in 1947. He flew the Bell X-1 with manual flight controls—unaided by computers—and electric motors to move control surfaces. SpaceShipOne also flew this way and became the first aircraft not built for a government-related program to fly faster than the speed of sound. *NASA*

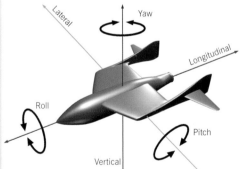

Just like normal aircraft, SpaceShipOne used flight controls to maneuver along the three primary flight axes—lateral, longitudinal, and vertical. SpaceShipOne did so by changing pitch and roll with elevons on the tail's horizontal stabilizers and yaw with rudders on the tail's vertical stabilizers. *James Linehan*

Once SpaceShipOne's rocket engine was lit, the pilot had to "turn the corner" right away. This maneuver allowed the energy the rocket engine to move SpaceShipOne upward as opposed to horizontally. The more time SpaceShipOne pointed vertically with its rocket engine roaring, the higher it would go. *Mojave Aerospace Ventures, LLC. SpaceShipOne, a Paul G. Allen Project*

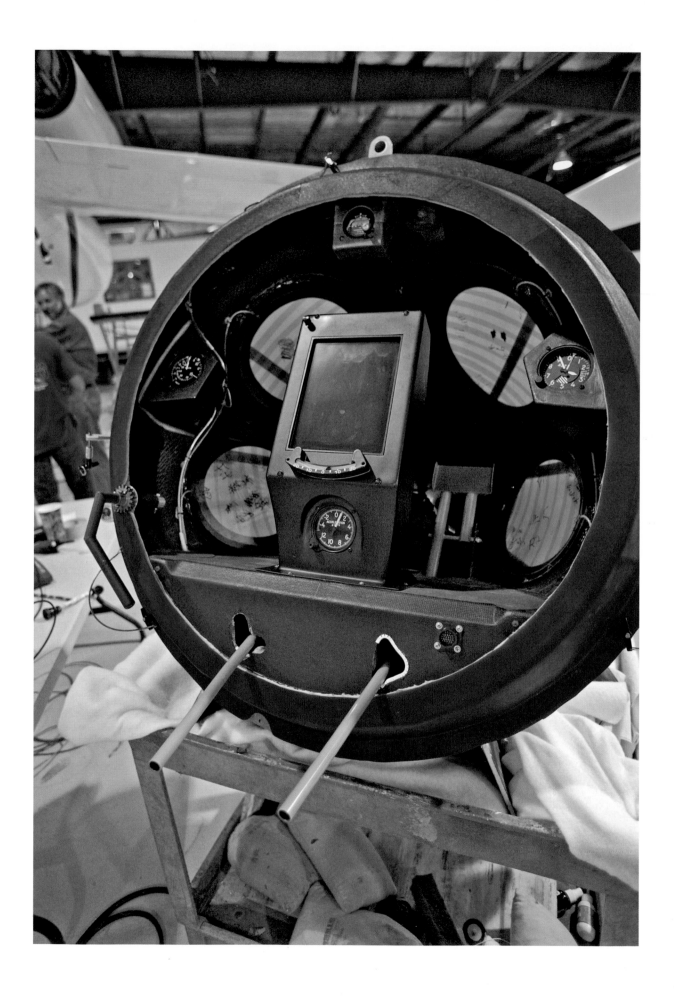

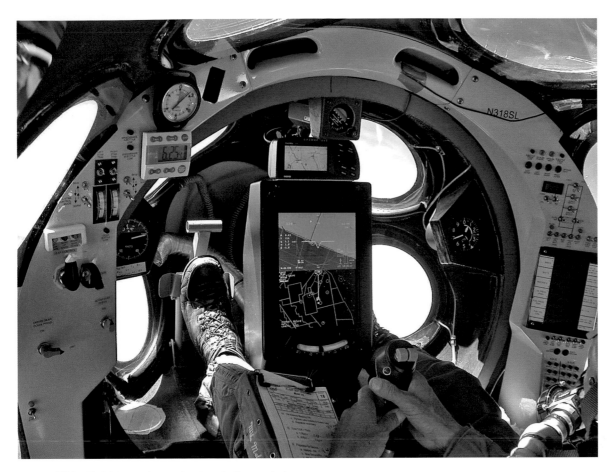

Above: White Knight, as shown, had a similar cockpit and instruments to SpaceShipOne. The vehicles were designed to keep as many components common as possible. This not only eased the manufacturing process and kept costs down, but since White Knight flew a year before SpaceShipOne, Scaled Composites could get good flight history on these common components. *Mojave Aerospace Ventures, LLC. SpaceShipOne, a Paul G. Allen Project*

Opposite: The flight director display (FDD), the rectangular video screen in the nose section, showed the readouts of the Tier One navigation unit (TONU). The SpaceShipOne pilot used the TONU to monitor important flight data and to help keep the spacecraft flying on a very precise trajectory. In the case of an emergency, the pilot could detach the nose section, as shown, and then parachute out. The two rods hanging at the bottom of the nose section are rudder linkages. *Mojave Aerospace Ventures, LLC. SpaceShipOne, a Paul G. Allen Project*

sixty miles per hour, it is a lot harder to keep your hand steady in the airflow. When you're traveling at six hundred miles per hour, as you put your hand outside, it would be sheered right off. That is the type of force a pilot must use his muscles to steer SpaceShipOne using a mechanical system, which is impossible.

So as SpaceShipOne got faster and approached the speed of sound, and the air pressure was pushing and pushing and pushing, the pilot then turned to an electric trim system. This was basically three electric motors that adjusted the trim rudders and horizontal stabilizers instead of the pilot trying to fight back and forth.

These electric motors were powerful enough to move the control surfaces, such as the stabilizers and rudders, in the face of the very high-pressure air that SpaceShipOne moved through. But after the spacecraft made its turn upwards and continued to head to space, the atmosphere rapidly disappeared. If the atmosphere disappeared, then the amount of air, or the number of air molecules, that the spacecraft encountered was less. Less air meant less air pressure and thus less force.

So as the spacecraft streaked upward, although it moved faster than ever before, there was actually less force acting on it. In order for the control surfaces to work, they needed to have air going over them because each of these control surfaces worked like little wings. Without air, these control surfaces could not generate the force needed to change or maintain course. As SpaceShipOne got closer and closer to space, it now was unable to use the mechanical system or the electric trim system to control its position, orientation, or direction.

What was needed then was a reaction control system (RCS), a third flight control system. Reaction control systems are used in all spacecraft that need to maneuver in space. They essentially are a series of small rocket engines.

SpaceShipOne had nozzles that released pressurized air. Just imagine the simple rocket engine created by blowing up a balloon and letting it go. The balloon suddenly flies off until the air inside runs out. It is the air going from the inside to the outside that provided a small amount of thrust, causing the balloon to go zipping along. SpaceShipOne's reaction control system behaved just like that. But since there is no sound in the vacuum of space, you wouldn't hear the air blowing outside the spaceship.

Squirting puffs of air certainly does not create very much thrust compared to more powerful chemical rocket engines. But Scale Composites could avoid a whole separate kind of propellant and its associated systems and redundancies by using the same gas that lifted and lowered the feather and activated other various components. So this greatly simplified the design.

SpaceShipOne had a series of nozzles that pointed in different directions. By switching the ports on and

off, the pilot could change where SpaceShipOne pointed while in space.

Typically, space vehicles do not use compressed air because compressed air coming out of a port is an inefficient rocket engine. Because of their mission requirements, the Space Shuttle, *International Space Station*, and satellites in orbit use very efficient propellants that produce a lot of thrust compared to their propellant's weight.

The Space Shuttle must rendezvous and dock with the *International Space Station*. The *International Space Station* must sometimes maneuver to avoid deadly space debris. Satellites have to maintain proper orbit and orientation or they will become nonfunctional.

This is where the KISS principle makes yet another big difference. If you already have a very large supply of propellant, in this case compressed air, and maneuvering in space isn't critical to the mission, then trying to develop a complex reaction control system where you have to buy expensive chemicals is simply overkill.

The fourth flight control system was the feather, which has been discussed in detail previously. On reentry, the feather was required to get SpaceShipOne through safely. It controlled the flight path of SpaceShipOne as it transitioned from space back into the atmosphere.

If SpaceShipOne was spinning or tumbling or doing some other gyration in space and the pilot wanted to counteract it, then he would normally use the reaction control system. However, it wouldn't be crucial for him to correct these motions because unlike the Space Shuttle where the orientation and trajectory were critical on reentry, SpaceShipOne could be rolling or even reenter upside down. The feather would self-right SpaceShipOne, putting it into the proper orientation automatically.

So if the reaction control system failed altogether, SpaceShipOne could get around it. The reaction control system did not have to be very complicated, but without the feather, SpaceShipOne faced damage during reentry.

TONU

SpaceShipOne used a computerized system called the Tier One Navigational Unit, or TONU. It was the main instrument the test pilot used to get all the readouts and trajectories and speeds and other flight data needed to fly the mission. This information was presented on one simple computer screen, which automatically stepped through screens for the various stages of the flight. Only the details that the test pilot needed for that particular stage of the flight were displayed, such as firing the rocket engine during boost, floating in space using the reaction control system, or gliding to the runway for landing.

However, the TONU didn't help fly the aircraft for the test pilot. If the TONU died, then the test pilot had backup instruments for the basic flight data. The backups resembled what would typically be seen in more conventional aircraft—gauges, meters, and dials.

Common Components and Construction

As the design of SpaceShipOne started to finalize, Burt Rutan could now go back and look at the vehicle that would eventually carry SpaceShipOne on the first part on its journey to space—White Knight. At first glance, White Knight and SpaceShipOne don't look anything alike. However, looks can often be deceiving, especially when you cannot see underneath the hood.

The nose and the cabin of the vehicles are nearly identical as well as the front part of the fuselage. Because of this common shape, two completely different vehicles didn't have to be developed, so this made the fabrication process quicker and cheaper in places. Both vehicles were made using a carbon fiber, epoxy, and honeycomb composite sandwich structure.

Since these common assemblies carried the crew for each vehicle, having two of a kind gave an extra degree of practice and confidence when it came to building the assembly for SpaceShipOne, which would eventually undergo the extremes of space travel.

SpaceShipOne was rocket powered and White Knight was jet powered, so they had a different set of instruments and controls for each. But many of the other components that went into building each vehicle were the same, such as the doors, windows, fittings, mounts, cockpit consoles, environment systems, and all kinds of other guts.

In the early stage of construction, White Knight's fuselage was assembled by bonding together large composite sections. These sections were shells made of carbon fiber/epoxy composites. When completed, the fuselage would have a strong, lightweight honeycomb structure. The forward section of SpaceShipOne's fuselage is also shown below White Knight. *Mojave Aerospace Ventures, LLC. SpaceShipOne, a Paul G. Allen Project*

Windshield

One thing that jumped out at first glance of SpaceShipOne was its windshield made up of sixteen round windows. There was actually a very important reason for this design, besides the somewhat cool and futuristic aesthetic. Having a large, single-piece windscreen would not just be heavy, but it would also reduce the overall structural strength. This type of windscreen would need to have strong supports to hold it in place. Also, each window had to be totally redundant—an inner and a separate outer window pane, each strong enough to handle the pressure if one failed.

A large windscreen that is redundant would be very heavy and a nonround shape would also be much heavier. With an array of circular windows, more of the fuselage, by area, was then the stronger and lighter composite material, compared to weaker and heavier window material. These circular windows were spaced to give the pilots a specific attitude reference—the top nose windows aligned with the horizon during the glide home, the bottom nose windows aligned with the horizon at touchdown on the runway, and four other windows aligned with the horizon during the supersonic portion of the rocket boost. Aside from reducing the weight and improving the strength of SpaceShipOne, the design opened up a maximum of viewable area.

Although SpaceShipOne had a 60-mile glide range after reentering, it was not a high performing glider. The pilot would have to carefully fly the landing approach. SpaceShipOne didn't have control surfaces to slow it down for landing. It could drop its landing gear, but that was a one-time maneuver. Once down, the landing gear stayed down until retracted by the ground crew. *Mojave Aerospace Ventures, LLC. SpaceShipOne, a Paul G. Allen Project*

By getting the mothership flying first, it acted as a testbed that enabled Scaled Composites to check out the functionality of the common components. Shaking out these on the vehicle under conditions less critical and more forgiving was much less risky than trying to shake them out in the vehicle that was going to break the sound barrier, go to space, and reenter the atmosphere.

Like any mechanical system, there is a period of breaking in and seasoning. This not only gives confidence that all this new stuff is doing the job it is supposed to do but shows reliability over the long run.

And sometimes a part used in White Knight at one place would be used in SpaceShipOne in a different place, as with the air-actuated cylinders used for the landing gear on White Knight. The same type of cylinders was used to raise and lower the feather.

The feather design was such an absolutely crucial part of SpaceShipOne's design. Again, if the feather failed to operate while on the way back from space, SpaceShipOne faced damage on reentry. Using White Knight to test out parts gave SpaceShipOne a great head start and significantly reduced the level of risk.

Right: The head-on view of White Knight and SpaceShipOne appears to show more differences than similarities. In fact, White Knight was designed to be as similar as possible to SpaceShipOne. White Knight shared many of the same components and systems, so it could qualify them prior to SpaceShipOne even flying. White Knight could also duplicate some of SpaceShipOne's flight characteristics, which also allowed the pilots to use it as a trainer.
Tyson V. Rininger

Right: A honeycomb sandwich structure was used to build the cabin walls of SpaceShipOne and White Knight. Outer and inner carbon fiber composite shells were separated by Nomex honeycomb structures. And each porthole had an outer and inner window. The SpaceShipOne crew did not need to wear spacesuits since the cabin was pressurized and the duration of the flight was so short. However, oxygen masks were worn as a precaution. *Mojave Aerospace Ventures, LLC. SpaceShipOne, a Paul G. Allen Project*

Nose Skid

Another highly effective and simple solution was the use of the nose skid. Landing skids have been used in space vehicles before. The X-15, for example, used a landing skid at the rear of the vehicle, not the nose as with SpaceShipOne.

There is a significant difference between an aircraft with retractable landing gear and an aircraft with fixed landing gear. Fixed landing gear is a very simple system. The landing gear is always out, and you never have to worry about it getting stuck in the up position or being stuck in the down position. The weight of the whole fixed landing gear system is much less than the weight of a retractable landing gear system.

The advantage of having retractable landing gear is that your performance is much greater. You don't have these huge objects dangling from the aircraft creating lots of drag. A landing skid is just about in the middle of these two types. And for SpaceShipOne, it was really the best of both worlds in a way. It took both the advantages of fixed and the retractable landing gear systems.

By simply deciding not to have a wheel and replacing it with a landing skid, you save an enormous amount of space as well. And once you have reduced the footprint, you can now easily make it retractable.

SpaceShipOne's nose skid was light, thin, and curved to fit the outer contour of the fuselage. This made it very lightweight, and because the landing skid only went one way, there was no internal retraction mechanism. It was spring loaded. So before a flight, the nose skid was reset by the guys in the hangar. The nose skid was also designed in such a way that it acted as a brake and a shock absorber without any additional moveable parts.

"The skid provides reliable braking, so you will decelerate on the runway even if your wheel brakes fail," Rutan said. "So if you only land and never take off, it is fine to have your brakes on all the time."

After SpaceShipOne came to stop on the runway, a pickup truck towed it away. This taxi system was cheap, reliable, and external to SpaceShipOne.

Parts and systems that added weight and could have a failure were eliminated. So for a potentially complicated system—such as front landing gear, a critical system because SpaceShipOne would crash if it failed on landing—Rutan distilled it to probably the simplest form possible. This improved performance, reliability, and cost effectiveness. The nose skid is one of the single best examples of an elegant design in terms of bang for the buck.

Shown here after returning from space, SpaceShipOne got a tow off the runway from Paul Allen, Burt Rutan, and Richard Branson (front to back). Because SpaceShipOne had a nose skid, it could not taxi by itself. The nose skid epitomized elegant design. It acted as a brake, supported the vehicle, was lightweight, didn't take up a lot of room, was easy to build, and didn't cost a lot. *Dan Linehan*

White Knight

Proteus set a pretty high standard for a wild look as far as aircraft go. So when White Knight rolled out, there was no surprise to see such a surprising looking aircraft, which resembled a spindly looking bat or a sci-fi spaceship. Rutan had already designed SpaceShipOne, but it wasn't built yet. Now he was ready to design the first ever aircraft that had the sole purpose of launching a spaceship.

With twin booms and twin tails, the configuration was never flown before on a manned jet aircraft. The goal for building White Knight was to build it as quickly and inexpensively as possible. It was built around the requirements for launching SpaceShipOne and around a powerful set of jet engines.

White Knight proved to be one of Scaled Composites' best handling aircraft from the start of flight testing. But like any aircraft in development, there were tweaks. Spoilers on the top of the wings were disabled and bolted down. Angled-up wingtips were installed to improve flying qualities.

"We used J85 engines off a Northrop T-38 because they were cheap, and we essentially got them off eBay," said Doug Shane, who piloted the first flight. "They were probably the worst-suited engines for the airplane that we could have selected, except for the fact, as I mentioned, they were cheap."

White Knight was used as a trainer for SpaceShipOne as well. The mothership also used a TONU, and since the cockpits were nearly identical, the test pilots could get a great degree of familiarity with SpaceShipOne even before sitting inside it. For safety, the door and emergency hatch were in the same place.

Right: Equipped with two J85-GE-5 turbojet engines, White Knight lifted SpaceShipOne to a launch altitude of about 47,000 feet (14,330 meters). To reach that height, White Knight would have to engage the engines' afterburners. Though not the best engines for the application because flameouts and dropping out of afterburners were not uncommon, they were cheap and still got the job done. *Mojave Aerospace Ventures, LLC. SpaceShipOne, a Paul G. Allen Project*

These two strangely wild-looking aircraft were both designed to be spaceship launchers. Proteus, flying below White Knight, was to be used for Burt Rutan's original spaceship concept, which was a single-place rocket. However, because of the size of three-place SpaceShipOne, it was necessary for him to build a bigger launch vehicle. *Mojave Aerospace Ventures, LLC. SpaceShipOne, a Paul G. Allen Project*

White Knight Details

Model number	318
Type	high-altitude, utility
Prototype tail number	N318SL
Current prototype location	in operation with Scaled Composites
Customer	Mojave Aerospace Ventures
Fabrication	Scaled Composites
Flight testing	Scaled Composites
First flight date	1 August 2002
First flight pilot	Doug Shane (pilot) and Pete Siebold (flight engineer)
Seating	one pilot (front seat) and two passengers (back seat), space-qualified, pressurized cabin
Wingspan	82 ft
Wing area	468 ft^2
Fuselage diameter	60" (maximum outer diameter)
Payload capacity	8,000-9,000 lbs
Gross weight	19,000 lbs (at takeoff with SpaceShipOne)
Engine	two J85-GE-5 turbojets with afterburners, 7,700 lbs total thrust
Landing gear	two fixed front and two retractable rear
Fuel capacity	6,400 lbs
Fuel type	JP-1
Never exceed speed (V_{NE})	0.6 Mach (160 KEAS)
Range	500 miles
Ceiling	53,000 ft

Test pilots had much more practice and experience extricating themselves in case of an emergency.

The powerful jet engines, control surfaces, and lightweight, aerodynamic shape of White Knight also enabled the test pilots to simulate certain flight characteristics of SpaceShipOne, similar to how the shuttle training aircraft—Gulfstream jets modified to handle like the Space Shuttle in glide—allowed NASA astronauts to practice gliding and landing.

The lightweight composite structure of White Knight allowed for a huge thrust-to-weight ratio. This let the pilot simulate the SpaceShipOne boost profile. However, the jet engines and their afterburners functioned poorly when reaching altitudes around fifty thousand feet.

Inboard and outboard speed brakes, when deployed, allowed the pilot to also simulate the glide of SpaceShipOne.

By the eighth flight test, White Knight had qualified the entire envelope required for launching SpaceShipOne. After flying additional test flights, four pilot certifications, and two air shows, White Knight flew SpaceShipOne into the sky for the first time. This was only White Knight's twenty-fourth flight.

After SpaceShipOne flew its last flight, Scaled Composites made White Knight available for other missions, such as "reconnaissance, surveillance, atmospheric research, data relay, telecommunications, imaging, and booster launch for microsatellites."

White Knight was one of the most important components of Scaled Composites' space program. Shown here returning after launching SpaceShipOne for its first spaceflight, White Knight enabled Scaled Composites to build SpaceShipOne safer, lighter, and cheaper by lifting SpaceShipOne above most of Earth's atmosphere before a rocket engine was ever lit. *Tyson V. Rininger*

Flights and Spaceflights

Flight tests of SpaceShipOne began 20 May 2003. Scaled Composites started with a captive carry mission, where White Knight simply lifted SpaceShipOne into the sky and flew it around. Scaled Composites needed to figure out the basics. For example, is this going to couple together without any adverse interaction? Imagine when you strap something to the roof of your car—like a canoe or mattress—and you drive highway speeds. If you don't have it tied down right, then the thing could easily rip off your car and create a whole lot of problems.

In a way, those at Scaled Composites were just making sure that SpaceShipOne and White Knight could fly together without creating problems for each other. SpaceShipOne went up this way unmanned. The next incremental step was to put a pilot inside to begin the process of checking out systems while in flight. The pilot moved the controls and saw how the forces built up. It was essentially like a mini wind tunnel but in situ.

After gaining confidence and everything checked out so far, the next step was to drop SpaceShipOne for a glide test with a pilot inside. The pilot would be able to fly freely and see how everything worked and then just glide home. Mike Melvill described this as one of the scariest flights he had done in the program because of the unknowns. SpaceShipOne had never been in the air by itself.

Typically, when Melvill had done the first flight of a vehicle, it would start off slowly with taxi tests on the runway, just to have air flowing over the airplane. He would get an initial feeling of its flying qualities. After that, he would lift a few feet off the runway to see that nothing squirrely happened and there were no hints of dangerous handling issues.

In other words, he was starting out pretty close to the ground, whereas SpaceShipOne's first flight was at forty-seven thousand feet, nearly nine miles above ground. So it better well work. Otherwise, Melvill would have to use the escape hatch by unscrewing the nose of the cockpit and pushing it off and then jumping out.

Each subsequent glide flight pushed the flight envelope open more and more in terms of speed, g-force, mass, maneuvering, and altitude. Even the feather was raised and retracted in glide test flights. After the flight envelope expanded about as far as it could go by unpowered flights, on 17 December 2003, Brian Binnie flew the first rocket-powered flight test of SpaceShipOne. On this day, one hundred years earlier, Wilber and Orville Wright made the first powered and controlled flight of a heavier than air vehicle at Kitty Hawk, North Carolina.

Test pilots Pete Siebold, Mike Melvill, and Brian Binnie (front row, left to right) each flew SpaceShipOne missions. They also flew White Knight when it carried SpaceShipOne. SpaceShipOne designer Burt Rutan (left) and test flight director Doug Shane (right) stand behind the test pilots. In 2008, Rutan semiretired from Scaled Composites. Shane took over as president, and Siebold became the new flight test director. *Mojave Aerospace Ventures, LLC. SpaceShipOne, a Paul G. Allen Project*

SpaceShipOne Flights

Date	Flight Number	Mission	Pilot
5/20/03	01C	captive carry	unmanned
7/29/03	02C	captive carry	Mike Melvill
8/7/03	03C	captive carry	Mike Melvill
8/27/03	04GC	captive carry	Mike Melvill
8/27/03	05G	glide	Mike Melvill
9/23/03	06G	glide	Mike Melvill
10/17/03	07G	glide	Mike Melvill
11/14/03	08G	glide	Pete Siebold
11/19/03	09G	glide	Mike Melvill
12/4/03	10G	glide	Brian Binnie
12/17/03	11P	rocket powered	Brian Binnie
3/11/04	12G	glide	Pete Siebold
4/8/04	13P	rocket powered	Pete Siebold
5/13/04	14P	rocket powered	Mike Melvill
6/21/04	15P	spaceflight	Mike Melvill
9/29/04	16P - X1	spaceflight	Mike Melvill
10/4/04	17P - X2	spaceflight	Brian Binnie

Flight tests began as captive carries with SpaceShipOne slung underneath White Knight the entire time. It was, in fact, like a giant wind tunnel in the sky. These first test flights, unmanned and manned, made sure that the vehicles coupled together without problems. When manned, the test pilot could exercise some of the systems, gaining familiarity and getting the feel of them. *Tyson V. Rininger*

"When you light that rocket motor off, everything literally starts with a bang. There is so much energy associated with that rocket motor. It is like a tsunami sweeps through the cabin and literally takes you away," said Binnie.

"You really have nothing in your background or DNA to tell you that what is happening to you is good. You have no basis. Three or four seconds will go by, and you go, 'Ah, I'm not dead. Therefore, it must be going as they told me it was going to go.' "

The first rocket-powered flight test of SpaceShipOne was an extraordinary achievement for Scaled Composites. This was the company's first flight test of a manned vehicle to break the sound barrier.

After two more rocket-powered flight tests, each opening the flight envelope more and more, higher and faster, SpaceShipOne was ready to take a crack at space. And on 21 June 2004, Mike Melvill piloted SpaceShipOne to an altitude of 328,491 feet, only a few hundred feet above the requirement of the Ansari X Prize.

On this, only the fifteenth flight of SpaceShipOne, it reached space and became the first privately funded, built, and flown spacecraft ever to do so. Melvill earned the very first set of commercial astronaut wings. But for

all the importance of that spaceflight, the spacecraft was not configured to qualify as an Ansari X Prize winning attempt. In fact, SpaceShipOne was significantly lighter than what the Ansari X Prize required, where it had to be manned or ballasted for the weight of three people.

SpaceShipOne had just barely crossed into space. Going by the ratio that for every pound lighter SpaceShipOne was, it would go up an extra 150 feet higher, if Melvill had been a few pounds heavier, then it was quite possible that SpaceShipOne would not have reached the boundary of space.

Melvill wanted a dramatic way to demonstrate the effect of weightlessness. In the morning before this spaceflight, on the way to Scaled Composites, he stopped off and picked up some M&Ms to release in the cockpit of SpaceShipOne when he reached space. His choice of M&Ms was thoughtful. They were colorful, so they would show up nicely on the cockpit video. They were also hard-shelled, so he wouldn't have to worry about them melting all over the place after he released them. And then they were soft enough that if they got trapped somewhere where they shouldn't be, then they wouldn't jam things up and could easily be smashed. Thus, he proved that M&Ms melt inside your space helmet and not in your space capsule. But the most

There were eight glide test flights after the captive carry test flights. SpaceShipOne pilots would drop off White Knight at altitude. Although unpowered, these test flights opened up a significant amount of the flight envelope in terms of factors such as speed, altitude, g-force, stall characteristics, and feather performance. *Mojave Aerospace Ventures, LLC. SpaceShipOne, a Paul G. Allen Project*

After SpaceShipOne completed its glide test flights, the next step was to ignite the rocket engine. Three rocket-powered flights incrementally increased the burn time of the rocket engine from 15 seconds up to 55 seconds. With a longer burn time for each subsequent test flight, SpaceShipOne's max speed and max altitude also increased. Each of these test flights was still an envelope expansion. *Mojave Aerospace Ventures, LLC. SpaceShipOne, a Paul G. Allen Project*

important factor may have been that they shared his same initials.

To solve the problem that SpaceShipOne barely reached space even when much lighter than it would have been for an Ansari X Prize attempt, Scaled Composites first enhanced the rocket engine performance to give it more energy to help carry it higher. Next, Scaled Composites reduced the weight by removing as many extraneous things from the spaceship as possible, very much the way Charles Lindbergh scrimped and shaved anything possible to reduce the weight of the *Spirit of St. Louis*.

Lastly, when Melvill had flown the mission, he had encountered wind shear going up. He had difficulty maintaining a very good trajectory. Instead of pointing upward as the trajectory called for, he pointed more horizontally. So the energy was being applied downrange as opposed to being applied upwards toward space. That was a key factor in him nearly not making it to the altitude. Even with external influences, pilots had to fly a precise check trajectory. Only three months later, Scaled Composites had SpaceShipOne ready for its first Ansari X Prize attempt.

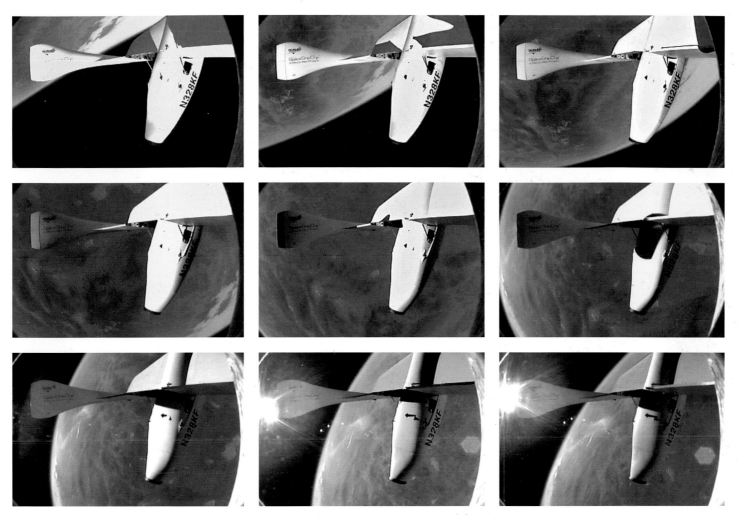

This sequence taken at three-second intervals shows the feather up as SpaceShipOne rotates in space. The feather was deployed while SpaceShipOne coasted to apogee even though the feather was needed for reentry. Since the feather was SpaceShipOne's way to safely return to Earth, the pilot wanted to get the feather up as soon as possible in order to troubleshoot any potential problem before SpaceShipOne started falling back to Earth. SpaceShipOne actually hit its fastest speeds on reentry. *Mojave Aerospace Ventures, LLC. SpaceShipOne, a Paul G. Allen Project*

The First Ansari X Prize Attempt

Up to now nothing went perfectly smoothly for SpaceShipOne. Flight test programs very rarely do. If they did, then there would be no need for them. There were failures and malfunctions and unexpected problems that led to design modifications. But that was the entire purpose of flight testing, to test how all the systems integrated with one another. You cannot expect the testing to go perfectly. That's just not realistic. SpaceShipOne even had a crash landing during its first rocket-powered flight test. On SpaceShipOne's very last test flight, it just made it to the level of space.

And now it was time to go for the prize. Scaled Composites had just one vehicle to make two spaceflights in only two weeks.

Scaled Composites normally does not publicize its test flights—for good reason. But Melvill's flight attempt to space had been a little bit of an exception.

Though it was an absolute milestone in the history of aviation, it still hadn't completely caught on to the rest of the world yet that all this was happening. That flight had showed a nongovernment program could indeed get to space.

When it was time for SpaceShipOne to take a shot at the Ansari X Prize, the world was certainly watching. Millions of people tuned in from across the globe and tens of thousands showed up in Mojave to watch the first attempt, which was called X1. So there was an extraordinary amount of focus. And there were no guarantees. The vehicle had to be significantly heavier because of the rules of the Ansari X Prize. Anything could go wrong. It was flying to space after all. It was an experimental vehicle, a completely new design. And a new type of rocket engine was used for the first time as well as a new type of reentry system. SpaceShipOne had to overcome all of these challenges.

Mike Melvill would pilot X1, and he was determined to make up for the poor trajectory during the first spaceflight. Now, as SpaceShipOne ascended in boost, in order to make sure the trajectory was steep enough, Melvill pulled more aggressively up and then found that he needed to push and trim nose-down to correct the trajectory. Thus, for the first time, SpaceShipOne was supersonic at negative lift, where the directional stability was worse than predicted. This caused a rolling departure. As the spaceship zipped up faster and faster and farther and farther out of the thick atmosphere, SpaceShipOne had lost control. It spiraled up, making twenty-nine rotations. He couldn't prevent it.

However, SpaceShipOne was not in a dangerous condition. Because of the rarefied atmosphere, there were not a lot of forces on the vehicle. Melvill didn't have to worry about SpaceShipOne disintegrating since he was high enough for it not to be a problem. So as SpaceShipOne spun, it fortunately spun in a manageable way. Melvill just held on, did not look out the windows, and let SpaceShipOne spiral up to space. Once up there, he used the reaction control system to counteract the rolling.

If Melvill aborted when the rolling started, he would have not made the Ansari X Prize altitude. This was why you have amazing pilots doing the job. They are able to judge a situation that looks pretty dire to most people on the ground, assess the problems, and can come up with solutions. To go or not to go? That's the test pilot's question, and the answer must come in a split second's time.

The reentry system was also a fallback. If SpaceShipOne was still rolling upon reentry, the feather system would eliminate the rolling because of how effective it was at slowing the spacecraft down and orienting it belly first. So even if he had run out of propellant for the reaction control system, he was still safe.

Scaled Composites had a very big problem, though. To most of the world it looked like, *Wow, these guys are out of hand. How is this going to be the start of commercial space travel if this is what's going on? Corkscrewing up into space is not the way to go. It's not on the flight plan.*

Rutan and the engineers had to figure out, and very quickly, what the problem was. And they did. They isolated the problem down to the fact that Melvill had flown SpaceShipOne in a regime that it was never flown before. The air was very thin when SpaceShipOne started rolling. Melvill needed to correct the rolling tendency that was happening because of asymmetric thrust coming out of the rocket engine. The control surfaces didn't have enough air flowing over them and had lost effectiveness to compensate for the roll as SpaceShipOne pointed nearly straight up.

The solution turned out to be a very simple one.

"For 17P we merely limited the amount of allowable 'down pitch trim', so Brian would definitely avoid the negative-lift condition that caused the departure," Burt Rutan said. "The solution was to more gently turn the corner, such that a forward correction later would not be needed. Pointing straight up at burnout is okay, as long as you do not push to negative lift. This quirk is fixed on SpaceShipTwo."

The entire spaceflight lasted about 24 minutes, and that's the time from when SpaceShipOne first dropped off of White Knight to the time SpaceShipOne landed on the runway. For about 3.5 minutes of that time, the pilot got to enjoy weightlessness. To demonstrate this effect, Mike Melvill released handfuls of M&Ms into the cockpit of SpaceShipOne while in space. *Mojave Aerospace Ventures, LLC. SpaceShipOne, a Paul G. Allen Project*

Opposite: This montage is actually a collection of images taken from a mobile telescope. Each image is a separate photograph that stitched all together shows the complete flight path of SpaceShipOne—ascending in captive carry, dropping off White Knight, boosting to space, coasting to apogee, reentering the atmosphere with the feather, and gliding back to Mojave Air & Space Port. *Ron Dantowitz, Clay Center Observatory, Dexter and Southfield Schools*

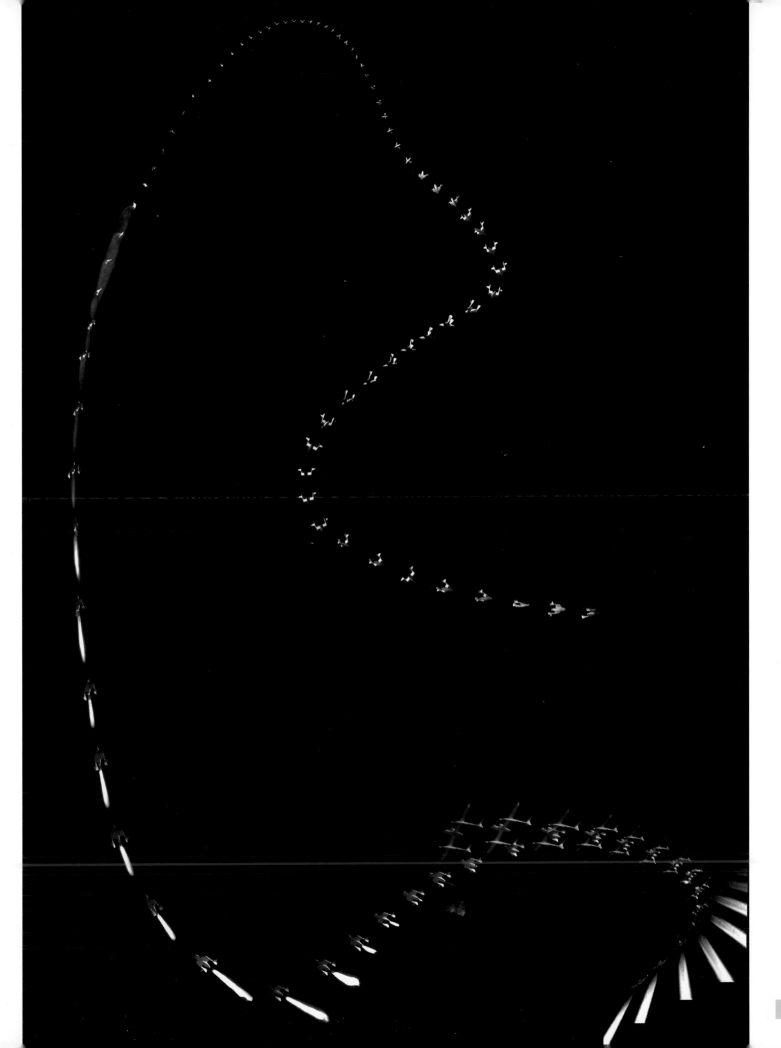

X2

Brian Binnie would be slated to fly the next spaceflight. Although Scaled Composites had a two-week window to make both flights, in only five days SpaceShipOne was ready to go back up again.

This test flight, 17P, was by far the most important flight of them all in terms of the ramifications for commercial space travel. It was only the seventeenth flight of SpaceShipOne. This was the $10 million flight. Make this and claim the Ansari X Prize. That was an awful lot of pressure. The prize expired in only a few months. Scaled Composites wouldn't have much time to redo it if there was a major mishap. It had to be done right.

The rules did not stop a team from flying a third time in that two-week window if the second attempt wasn't successful. The attempts did not have to be consecutive. But no one wanted to be in that situation. A third try would be a huge step back. Considering the corkscrew into space during the first attempt, if the second attempt failed, then that would certainly cast severe doubt on the viability of the program and on all commercial space travel. It was critical for SpaceShipOne to make it on the second try. SpaceShipOne had yet to fly a perfect and problem-free flight. Now, $10 million was on the line.

"I can safely say the one [thing] that made the pilots uniformly uncomfortable was the hour-long wait in SpaceShipOne while the White Knight carrier aircraft dragged it up to release altitude," said Binnie in an article he wrote for *Air & Space*. "During this time, there is little to do and the mind is somewhat free to wander."

Waiting an hour between takeoff and release from White Knight was more than torturous for Binnie. He couldn't wait to light the fire and stop being a passenger on the mothership. He wanted to take control. His last flight in SpaceShipOne had been ten months ago. It ended in a crash landing.

"For me personally, a problem or failure or inability to not pull this off for whatever reason, the other side of that coin was a bottomless pit. It felt to me like an abyss."

When Brian Binnie flipped the switch to fire the rocket engine, SpaceShipOne rocketed past so close to White Knight, which flew on blaring afterburners, that Mike Melvill and Matt Stinemetze could hear it from inside.

Scaled Composites now understood the rolling behavior of SpaceShipOne and figured out how to stay out of those conditions. But the truth was that during the previous flight, those at Scaled Composites thought they knew everything before and thought SpaceShipOne would fly perfectly fine during Melvill's spaceflight as well. But something new did come up then. Was there something else now waiting for Binnie, somewhere in a part of the envelope that hadn't yet been explored?

This was by no means a gimme flight by any stretch of the imagination. Since SpaceShipOne cut its engines halfway to apogee, the only way to know how high it would go was to follow an altitude-predicting instrument. It read 328,000 feet. Then it read 350,000 feet. Binnie gave the rocket engine a few more seconds of juice after that just for good measure. Now the world waited as SpaceShipOne coasted to space as the gap between true altitude and predicted altitude narrowed down.

"Hey, we're going to the stars. This is so cool," Rutan said of his feeling as he watched SpaceShipOne cruise past a true altitude of 328,000 feet, still with plenty of momentum.

SpaceShipOne hit apogee at 367,500 feet, which was not only well above the Ansari X Prize requirement but also above the previous record of 354,200 feet held by the X-15. Binnie had captured the trajectory and flew the first and only flawless flight of SpaceShipOne, a $10 million performance.

With just five days between spaceflights X1 and X2 of SpaceShipOne, Burt Rutan, Paul Allen, Mike Melvill, Brian Binnie, and the rest of the team from Scaled Composites captured the Ansari X Prize. Visionaries Peter Diamandis and Anousheh Ansari celebrated the victory of the first step of new space and were joined by Richard Branson, who stood ready to start the second step. *X PRIZE Foundation*

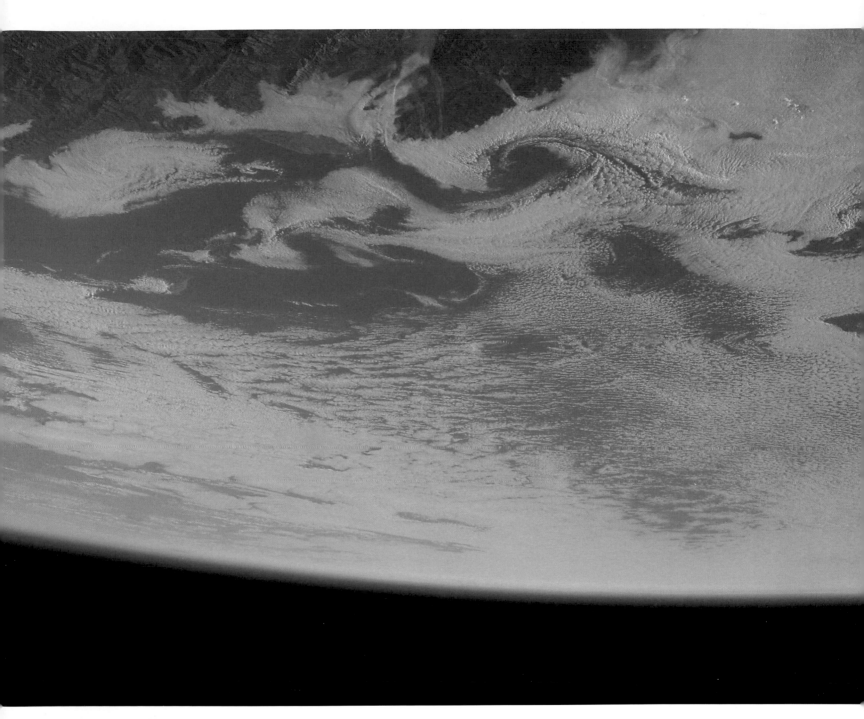

Brian Binnie photographed this view of Earth from inside the cockpit of SpaceShipOne on 4 October 2004. For the second Ansari X Prize attempt, he had just flown a perfect trajectory exiting the atmosphere and entering space. He reached an apogee of 367,500 feet, which was not only well above the 328,000-foot threshold of space but exceeded the previous world record of 354,200 feet, held by the X-15, by more than two and half miles. *Mojave Aerospace Ventures, LLC. SpaceShipOne, a Paul G. Allen Project*

A starship follows a spaceship. Or is it the other way around? Starship was the first aircraft Scaled Composites worked on in 1982. SpaceShipTwo, carried by WhiteKnightTwo, began its flight testing in 2010. Nearly three decades have passed between the two. Just as Starship stretched the boundaries way back then, SpaceShipTwo is prepared to do same. *Virgin Galactic/Mark Greenberg*

Chapter 5

SpaceShipTwo and WhiteKnightTwo: The Next Generation

Paul Allen and Richard Branson sat with Burt Rutan on the couch in his Mojave pyramid house in 2004. "What do you think can happen during your lifetime?" Rutan asked of them about space. He wondered what they thought they'd eventually be able to experience and enjoy. He felt that this question planted a stick in the ground and set a goal.

They spoke about dreams and desires that reached back to their childhoods. The high-tech philanthropist, the adventurous entrepreneur, and the aviation innovator sought answers. They pondered the future.

"I would love to play racquetball and golf in zero g," Burt Rutan said. "If you were at your resort hotel in Earth orbit, you could board a very simple spacecraft, which is nothing but an inflated cabin with some propulsion and fuel, that leaves Earth orbit and takes that beautiful swing around the Moon and then decelerates back into orbit. It doesn't have to have a wing or landing gear or thermal protection. It's a very simple spacecraft. But it gives people the view that Apollo 8 had."

Rutan had planted his stick. Humans had first set foot on the Moon in 1969—thirty-five years previous

to this lofty conversation. Now several years later, he remains convinced that air-launched spaceships and commercial suborbital space travel will eventually lead to "unbelievably cheap orbital space travel" and space hotels. If this is the case, then lunar excursions, like his wish, would not be that far off.

Reapplying a familiar classification system, Rutan designated suborbital projects as Tier One, as with SpaceShipOne; orbital projects as Tier Two; and projects that reached other bodies in space, like the Moon, as Tier Three. However, before reaching Tier Two and Tier Three, a big step was needed. By building on the technology Rutan and Allen developed, Branson would begin that step, which happens to be called Tier One-B.

Illustrated by Cory Bird, what looks like a scene from a 1950s Disney cartoon or a sci-fi classic like *2001: A Space Odyssey*, space stations and space hotels not too different than these have already begun to be developed. And as spacelines and spacecraft continue to grow in capabilities, it won't be too far off that people will be able to take a trip into orbit to check into their space hotel rooms. *Courtesy of Burt Rutan*

Sir Richard Branson on Space and Spaceships

How did you get to know Burt Rutan before SpaceShipOne came along?

I first met Burt Rutan in the early 1990s when he was building a carbon composite capsule for a global balloon project called Earthwinds. It was then I first learned about the amazing work he was doing in aerospace with new-technology materials. We then met again in 1999 when my colleague Will Whitehorn and I were looking at a private space project in Mojave, the Rotary Rocket, which we decided not to invest in. A couple of years later, my friend Steve Fossett and I jointly funded Burt to build a very special aircraft called the Virgin Atlantic Global Flyer. It was at that point that Virgin's engineers, marketers, and ideas people got their first real exposure to Burt and Scaled Composites.

How did you first find out about SpaceShipOne?

Will Whitehorn phoned me from the factory where he and another colleague, Alex Tai, were looking at the construction of the Global Flyer. It was a very excited phone call in which I found out that Burt was building a contender to win the X Prize. I already knew about the X Prize because Dr. Peter Diamandis had tried in the late 1990s to get Virgin to sponsor it. We had met him in London and the conclusion was that rather than sponsor the prize we would consider investing in whoever won it.

What made you think that Scaled Composites could build a spaceship for you?

Quite simply the fact that SpaceShipOne looked like something that could be developed into a commercial space vehicle. The fact that Burt is a genius and that Scaled is the undisputed master of composite materials also helps. I think our team had also learned a lot about Scaled from the Virgin Atlantic Global Flyer project, and they had a lot of confidence that Scaled could rise to the challenge of the much bigger Virgin Galactic project.

Why did you want to create a spaceline?

Tourism has always been at the forefront of the industrialization of transport, and that is true from shipping and the railways right through to early aviation. I have always wanted to go to space myself, and I had a gut feeling that many thousands of others would do the same or would feel the same way if access to space could be safe enough and cheap enough. I also believe that space tourism is the key to turning space from an extremely expensive monopoly of science and the military into an area much more widely used for commercial and industrial activity. Creating a successful spaceline is something that I believe could kick start the whole industry.

What excites you most about new space?

New space excites me because finally the private sector has been given the opportunity to do what it does best—innovate. The fact that space is ceasing to be a monopoly for big old-fashioned government rockets is a good thing. We now have the technology to put so much more into space than we do at the moment, if we can only lower the cost and increase the frequency and ease of access. There are now lots of entrepreneurs coming forward to do just that, and it's very exciting—people like Elon Musk, who is also a Virgin Galactic space customer and is doing wonderful work at SpaceX.

What are you most looking forward to on your first ride to space?

Most people tend to say that the thing they are most looking forward to about their ride into space with Virgin Galactic is the amazing experience of weightlessness. For me I think it is the thrill of the ride up, pulling all those g's. I have already experienced in the centrifuge. Followed by the silence of space, the prelude to that amazing view of the planet Earth, which I can't wait to see. Looking down on that curved Earth and that thin, thin layer of atmosphere, protecting billions of people from the vacuum of space, will be an awesome sight.

SpaceShipTwo

"I remember sitting on the sofa with my parents back in 1969, watching the incredible TV footage of the first Moon landing, and I knew that one day I wanted to go to space," Richard Branson said of this life-changing memory. Branson got his start in the rock 'n' roll business but in 1984 founded the UK-based Virgin Atlantic Airways. This airline has never had a fatal accident, and its growth helped allow its parent company to form other airlines in the United States and Australia.

Starting a spaceline was something that Branson had been thinking about long before Virgin Galactic was officially announced. Branson even visited Rotary Rocket to see Roton during a test flight.

Before Mike Melvill and Brian Binnie flew SpaceShipOne to space to capture the Ansari X Prize, Scaled Composites was approached by four different investors wanting to fund the next-generation spaceship. Branson was initially interested in being just a spacecraft operator, not a spacecraft builder. However, Branson wanted to do what he could to ensure the new space industry developed.

When it came down to make a decision on an investor, what clinched it for Rutan was the passion for space he saw in Branson's eyes and that this passion was genuine. Rutan knew there would be tough challenges ahead, but he also knew that Branson's commitment would be unwavering.

In 2001, Scaled Composites partnered with Paul Allen's Vulcan, forming Mojave Aerospace Ventures to conduct Tier One, the SpaceShipOne research test program. In 2006, Scaled Composites contracted with Branson's Virgin Group to conduct the first research program, Tier One-B, to develop a prototype and do the initial flight tests of the first commercial spaceship, SpaceShipTwo (SS2). Virgin also funded the startup of The Spaceship Company (TSC) to license and manufacture the SpaceShipTwos and WhiteKnightTwos (WK2) that will enter commercial service, flying the public to space. Founded by Branson in 1999, the spaceline Virgin Galactic will operate those spaceflights. An airline flies the public in airliners. A spaceline flies them in spaceships.

Airline companies, on the other hand, leave the development of airliners to manufacturers like Boeing

This time Richard Branson, Burt Rutan, and Paul Allen (left to right) are looking up to see SpaceShipOne descend from space after reaching the altitude required to win the Ansari X Prize. With SpaceShipTwo, they'll have the chance to see from the other side. *X PRIZE Foundation*

and Airbus, though these manufacturers will, of course, do their very best to make their customers happy. Airline companies are still free to fly airliners from any manufacturer. It will be interesting to see how Virgin Galactic will embrace spacecraft from other manufacturers as the market for commercial space travel grows. But commercial space travel is virgin territory, and the sky is no longer the limit.

Virgin Galactic will start with an initial fleet of five SpaceShipTwos and three WhiteKnightTwos. The Spaceship Company established new facilities at the Mojave Air & Space Port for the construction of the new SpaceShipTwos and WhiteKnightTwos that will round out Virgin Galactic's fleet. Only the prototypes were built directly by Scaled Composites, so it now has handed off the manufacturing to The Spaceship Company.

In 2009, Virgin Galactic significantly boosted its financial commitment to commercial space travel as it announced an exciting new partnership with Aabar Investments out of Abu Dhabi, United Arab Emirates. Aabar Investments put in $280 million, making the company a 32 percent shareholder of Virgin Galactic.

It is very encouraging to see partnerships forming beyond national boundaries. After all, borderlines are not viewable from space, and the view into space can been seen from anywhere on our planet. The Ansari X Prize hoped to achieve an international march into space. And very fortunately for the success of commercial space travel, this cooperation appears to be happening.

An Everyday Person's Spaceship

After SpaceShipOne won the Ansari X Prize, what happened to it? Scaled Composites still did have a few extra rocket engines on hand and made provisions for Task 21.

"Task 21 was that we would fly SpaceShipOne every Tuesday for five months, reasoning that if we did that you could then make with confidence a commercial business plan," Burt Rutan said.

But Task 21 wasn't funded. Rutan had figured that once he got the data together on the costs of the spaceflights that had flown, he would then approach Paul Allen.

"That would be the opportunity for Paul and me and both of our friends to be astronauts," Rutan explained. "If you just count only the passengers, you've got forty-four people. So maybe twenty of my friends could be astronauts and twenty of his friends could be astronauts. That would be kind of cool. That was the plan."

But something got in the way of the plan. Rutan underestimated SpaceShipOne's impact on the public, media, and historians. After SpaceShipOne's first spaceflight in June 2004, the Smithsonian's National Air and Space Museum thought that SpaceShipOne was an important artifact representing the future commercial spaceflight industry. The museum wanted to hang SpaceShipOne in its Milestones of Flight Gallery.

When that request came, Paul Allen didn't want to fly SpaceShipOne anymore after the Ansari X Prize.

WhiteKnightTwo Details

Model number	348
Type	commercial space vehicle launcher
Prototype tail number	N348MS
Customer	Virgin Galactic
Fabrication	Scaled Composites
Flight testing	Scaled Composites
First flight date	21 December 2008
First flight crew	Peter Siebold (pilot) and Clint Nichols (copilot)
Seating	2 crew, 14 passengers
Wingspan	140 ft
Wing area	1,315 ft^2
Tail height	25 ft
Length	78 ft
Gross weight	approx. 66,000 lbs
Payload	35,000 lbs
Engines	4 Pratt & Whitney Canada PW308A turbofans
Landing gear	quad retractable, front and back of each boom
Fuel capacity	22,000 lbs
Cruise speed	Mach 0.65
Ceiling	60,000+ ft

But Scaled Composites had three or four motors, so it could have flown more spaceflights even without funding Task 21. Rutan's first reaction was to argue why Scaled Composites should keep flying. He said, "You've got to prove a business plan. If this is going to go on to the next step, you got to do this." However, Rutan quickly realized that Allen was right in wanting to ensure SpaceShipOne was saved for history.

A Second SpaceShipOne

This book goes to print seven years after SpaceShipOne made its first spaceflight on 21 June 2004. It is interesting to think what would've happened if Scaled Composites used the tooling it had, built another spaceship just like SpaceShipOne, and started flying it to space commercially.

Consider asking people, "Would you like to take a ride to space?" I'd imagine people would fall into four basic groups based on their answers: (1) those who would never go up no matter what, (2) those who would take an awful lot of convincing but would end up going, (3) those who could stand to wait for a better ride, and (4) those who don't care that much about the ride as long as it gets them where they want to go.

I would be at the front of the line to volunteer as a spaceflight crash test dummy. So I figure that puts me squarely in the fourth group. The ride in SpaceShipOne No. 2 wouldn't be as high or as long or as unrestricted as in SpaceShipTwo, but when I talk about my trip to Antarctica, most people never even wonder what boat I took to get there.

SpaceShipOne No. 2 could have been generating a lot of cash immediately. It would have siphoned some of the attention from the skilled labor force working on SpaceShipTwo, but things didn't get started on SpaceShipTwo right away anyway. SpaceShipOne No. 2 could have been a good stepping stone for SpaceShipTwo by allowing those things that come up when starting something completely new to be addressed by a transitional vehicle—*we know we have this problem or that problem but it will be hammered out by the time the flagship is ready.*

SpaceShipOne No. 2 could also have had a lot of merit when considering the lull that has occurred since the fabulous and exciting times of the Ansari X Prize. Any spaceflight from then until now could have helped fill that void.

It seems that there was a window of opportunity that Scaled Composites could have capitalized on. After all, the Mercury astronauts, when strapped into their spaceship, had difficulty feeling the weightlessness. That wasn't a bad ride either. Regardless, making spaceships is a tough business. If it was well understood from the beginning how long it would have taken, maybe something like SpaceShipOne No. 2 would have been explored.

Homebuilt Spaceships

Imagine if you could build your own homebuilt spaceship. Burt Rutan got his start with homebuilts, right? Think of the kit: the fuselage from the guy who made the fuselages for Solitaire, the welded parts from the guy who made the welded parts for the Long-EZ, the other pieces from Spacecraft Spruce. I would sign any liability waiver in the world to get my hands on one of those. So, I wonder, who will be the first to design a homebuilt spacecraft? Burt?

How much would it cost? Design, development, construction, and testing of SpaceShipOne and White Knight, including the operation of three spaceflights and a few spare rocket engines, cost Scaled Composites $25 million. Imagine plunking down a few million dollars for a kit. You wouldn't have to build a mothership. With kits available, there wouldn't be any shortage of motherships for hire, just as glider pilots don't have to own a tow plane. Whereas the 1970s and 1980s were the era of canard homebuilts, what if the 2010s and 2020s become the era of the garage-built spaceships? What if not a fleet of VariEzes were parked out in front of Scaled Composites or at Oshkosh, but a fleet of SpaceShipOnes? Maybe it's too soon. It took a good seventy years after the Wright Flyer for the VariViggen kit to come out.

WhiteKnightTwo

"I have not been the designer, except for some basic concept stuff—only the guy who checks and advises changes—for the aircraft after SpaceShipOne and White Knight, which were designed in the 1999 to 2002 time period. Yes, the new folks are very good designers. I am very proud to have found them and to have them on our team here at Scaled," Burt Rutan said.

Rutan had just turned sixty-one when SpaceShipOne made it to space the first time. He had been burning the rocket engine from both sides for decades. In 2008, due to health issues, he stepped down as president of Scaled Composites. However, he is still involved with SpaceShipTwo and WhiteKnightTwo development, but not as the responsible designer. His role as mentor has become more important as he puts the company into the hands of the other fabulous engineers who fill the ranks in Scaled Composites.

After making some preliminary concept sketches, Rutan handed off the design responsibility to Jim Tighe, Matt Stinemetze, and Bob Morgan. They did major configuration changes and oversaw the team that did all the detail design of SpaceShipTwo and WhiteKnightTwo.

Build it Big

"The first thing we realized is that we couldn't rebuild SpaceShipOne," said Will Whitehorn, president of Virgin Galactic. "Customers are not going to pay two hundred thousand dollars to go to space and be cramped in a tiny cabin and not be able to move around and experience weightlessness and not be able to see the blackness of space and the brilliant planet below them."

So Burt Rutan sketched out a design with a fuselage centered on the wing, as White Knight's does, and two tail booms with vertical fins connected together at the top by a horizontal tail, similar to the tail of the ATTT (Model 133-B). It turned out that this tail configuration was complex and very difficult to analyze. Bob Morgan, WhiteKnightTwo project engineer and chief designer, recommended changing the tail configuration to something that more closely resembled that of White Knight and GlobalFlyer.

However, both these two vehicle had a single, centerline fuselage with the engines inboard as opposed to the twin booms and outboard engines of WhiteKnightTwo.

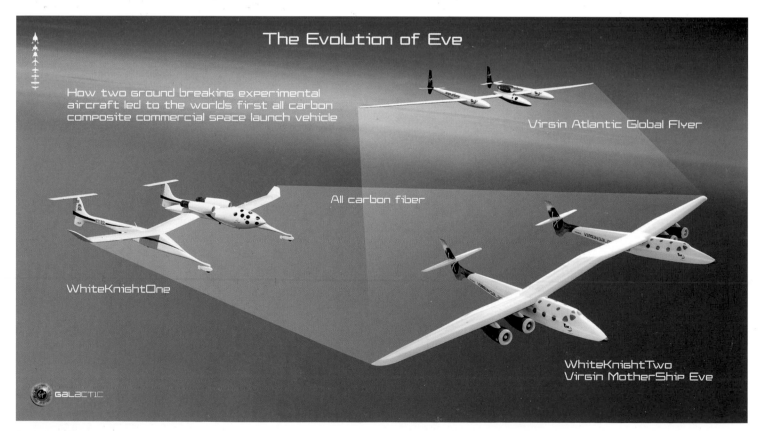

One doesn't have to look too hard to see the influence of White Knight and GlobalFlyer on WhiteKnightTwo. White Knight was built for hauling while GlobalFlyer was built for efficiency. To ensure commercial success, WhiteKnightTwo had to combine the best aspects of its two direct descendants. *Virgin Galactic*

During the SpaceShipOne spaceflights, Morgan got the idea that it would be wonderful to be able to watch the launch up close. But SpaceShipOne was slung underneath White Knight, and by the time it came into view after separation, SpaceShipOne was boosting to space nearly supersonic.

Morgan thought with twin booms, SpaceShipTwo would be in the middle, and with the cabins at exactly the same height, the passengers in WhiteKnightTwo could get a very good view of the launch. A host of other benefits resulted from having twin booms, such as improved aerodynamics, easier loading and unloading of passengers and flight crew, and a huge open space to carry lots of other exciting things besides SpaceShipTwo.

"Bob really came up with the justification that 'let's have the engines outboard the booms instead of inboard like White Knight,' " Rutan said. "And again the main reason was who knows what we were going to be carrying under that. You might want a spaceship with a V-tail, and it would be right in the engine plume."

Similar as with White Knight and SpaceShipOne, the new mothership has a cockpit and cabin nearly identical to the new spaceship.

"WhiteKnightTwo has the cabin structure, the windows, and the environmental control system of a spaceship," Rutan said. "We are qualifying that cabin to go to space. And everything that's intended to be in the spaceship to allow you to safely put people in space even without pressure suits is in this airplane."

At 7.5 feet (2.3 meters) in diameter, 12 feet (3.7 meter) in length, and able to carry six passengers, the cabin has a maximized volume so that when the passengers reach space in SpaceShipTwo, there is as much room as possible to float around in. Because of this, it was a bit of a challenge for engineers to fit all the systems into WhiteKnightTwo.

"It looks a lot different," said Rutan of the enormous WhiteKnightTwo. "It is not just a grown up version of White Knight."

It is not only the mothership that grew in size, though. Since winning the $10 million Ansari X Prize in 2004, Scaled Composites has doubled the number of its employees and its footprint as it gobbles up buildings along the flightline of the Mojave Air & Space Port.

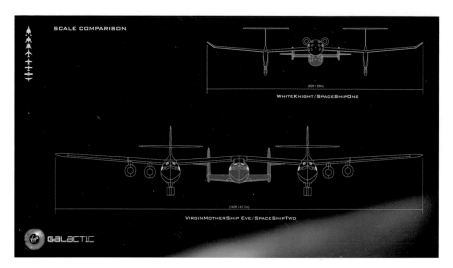

With a 49.5-foot (13-meter) clearance between the twin booms, there is an awful lot of room hanging under the giant wing of WhiteKnightTwo. This open architecture will not only allow WhiteKnightTwo to carry SpaceShipTwo but will provide the flexibility to carry many other payloads, some capable of reaching orbit. *Virgin Galactic*

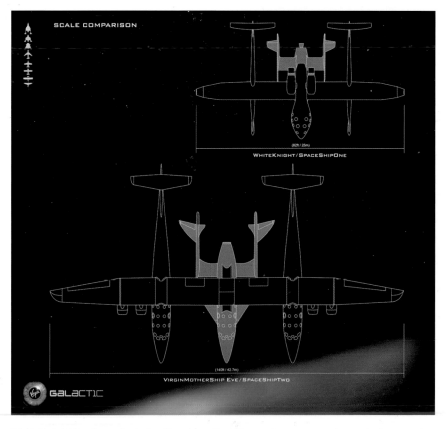

WhiteKnightTwo is really twice the aircraft of its predecessor and namesake—and sometimes a little more than that. The aircraft has two pilots, not just one. It has two cabins; the original has one. It has four engines compared to two. It has nearly twice the wingspan of White Knight. And WhiteKnightTwo is more than twice the launch weight, actually three times as much. *Virgin Galactic*

WhiteKnightTwo's wing and fuselages, or booms, are shown before being mated up. Two single-piece composite wing spars run nearly end-to-end inside the wing to provide support for the 140-foot-long wing and heavy payloads. These spars are the largest dimension composite parts, made of carbon fiber and epoxy, ever used in an aircraft. *Virgin Galactic*

Unveiling of *Eve*

On 28 July 2008, WhiteKnightTwo was officially unveiled. "This is a big moment for us," Branson said, having just climbed out of the cockpit with Rutan as the launch aircraft glistened in the sunlight like a star. "And I think you will agree that WhiteKnightTwo, or *Eve* as we will be able to officially call her soon, is one of the most beautiful and extraordinary aviation vehicles ever developed."

The smooth lines, the lovely curves, the silky transitions, and the two big booms—which of course are the fuselages that can both hold passengers—make WhiteKnightTwo a fine looking piece of aviation for sure.

Now christened the *Virgin Mother Ship (VMS) Eve*, the first WhiteKnightTwo is named in honor of Branson's mother, who was quite the aviation pioneer herself. Eve Branson tried to pose as a boy to get into Royal Air Force glider training during World War II. That was certainly a hard one to pull off, but nonetheless she was allowed to stay. However, after the war she couldn't fly an airliner because women weren't allowed. So she did the next best thing to keep herself in the air, which was to become one of the very first intercontinental flight attendants and fly aboard the de Havilland Comet, the very first jetliner.

A photograph of Eve as a young woman was used to create the sexy Galactic Girl insignia that will also grace the side of every Virgin Galactic mothership and spaceship.

WhiteKnightTwo will cruise along at Mach 0.65. And although Scaled Composites is shooting to have WhiteKnightTwo fly at high altitudes above sixty thousand feet, WhiteKnightTwo will launch SpaceShipTwo at around forty-eight thousand feet. This is above about 85 percent of the atmosphere and similar to the separation altitude of SpaceShipOne. Even though both SpaceShipTwo and WhiteKnightTwo are much bigger and more powerful than their first-generation counterparts, the air density at this altitude is still the most favorable to launch at.

After SpaceShipTwo reaches apogee, the only assistance it will have while plummeting back to Earth is that of gravity. Meanwhile, once free of SpaceShipTwo, WhiteKnightTwo will train the next group of astronauts, who will have just witnessed the SpaceShipTwo launch up close. WhiteKnightTwo then will fly a series of zero-g maneuvers. But unlike NASA's Vomit Comet, which porpoises in 0 to 2 g parabolas that each give zero-g periods of about twenty-five seconds, WhiteKnightTwo passengers float in zero-g for five to ten seconds and then WhiteKnightTwo takes another twenty-two seconds slowly building up to 1 g. This is to simulate the actual buildup rate of g-force felt when SpaceShipTwo enters the atmosphere after a spaceflight.

"Now while you do that, as you bring them down, you don't have to stop at 1 g or even 2 g, like the airliner does," Rutan said. "You roll and descend and pull and take them all the way to 6 g. So what you done is that you've taken passengers from floating all the way around to the maximum g-load they'll see on reentry."

So the passengers aboard WhiteKnightTwo get to practice over and over the same kind of g-force scenario they'll face during a spaceflight reentry.

The selection of the tail number is of special significance to Scaled Composites, and WhiteKnightTwo was given N348MS. All North American aircraft start with N, but the 348 stands for WhiteKnightTwo's model number and MS stands for mothership.

Internally at Scaled Composites, WhiteKnightTwo goes by a little bit of a different name. With SpaceShipTwo attached to WhiteKnightTwo, Bob Morgan noticed it resembled a triceratops, a three-horned dinosaur. Since White Knight is still in use and so as not to cause confusion between the two while talking about them, the name *T-Top* stuck for WhiteKnightTwo.

New Territory with Virgin

In all their years of operation, neither RAF or Scaled Composites built aircraft like a typical aircraft manufacturer. The aircraft they built were not mass produced and then sold to customers. The most complicated aircraft Scaled Composites ever built and made for commercial use, *T-Top* is a whole other beast entirely.

"All the primary structure is composites, except for the usual things like landing gear and the engine mounts," said Doug Shane. "Everything else is graphite/epoxy."

The two wing spars inside the 140-foot (42.7-meter) wing, which are the main structural elements of the wing, are the largest continuous pieces of composites ever used in an aircraft. They provide a wingspan larger than those of the Airbus A320 and Boeing 757. "The

A closeup of the giant tail is shown here during construction. A mechanical flight control system using pushrods and carbon fiber cables move the rudders and other flight control surfaces. No hydraulics or other power augmentation is used. WhiteKnightTwo does not have yaw dampeners or even an autopilot. *Virgin Galactic/ Thierry Boccon-Gibod*

Modern airliners have tended to go from four smaller engines to two larger engines in order to reduce cost and improve efficiency. But because of WhiteKnightTwo's configuration and the amount of weight it would be carrying, if it had only two engines and one engine ever went out on takeoff or landing, then it would be a very dangerous situation since its engines have to be so far from the centerline of the aircraft. As a safety measure four engines are used, two on each side. Shown here is one of WhiteKnightTwo's Pratt & Whitney Canada PW308A turbofan engines. This engine was the most fuel-efficient aircraft engine of its type available at the time of WhiteKnightTwo's construction. *Dan Linehan*

wing spar in this airplane has no secondary bonds," Rutan said. "It has no fasteners. It is full-span. . . . Composites don't bolt together very well, so we just don't use bolts at all."

To power the carrier aircraft, four Pratt & Whitney Canada PW308A turbofans are mounted in pairs outboard of each fuselage. "A key to this basic platform is its inherent redundancy in this configuration," said Pete Siebold, who flew SpaceShipOne three times and is now Scaled Composites' flight test director. "We can have an engine failure—even two engine failures—you won't be able to make the ultimate mission of dropping the spaceship, but at sea level and low altitude you still have excess performance in terms of climb capability and takeoff performance."

WhiteKnightTwo will also have range enough to ferry SpaceShipTwo coast-to-coast across the United States, which is significantly greater range than the five hundred miles of White Knight.

"Many of the Scaled test pilots are internally grown," Siebold said. "From Mike Melvill coming in as a machinist to work for Burt, being trained by Dick Rutan and Burt himself, to become a world-class test pilot and astronaut. And Doug Shane, the same. For his first job out of college, he came to Scaled and trained

with Burt and Mike and Dick as well. My first job out to school—I was a flight instructor—and I was trained as a flight engineer and subsequently a test pilot. We have a tradition of looking within to bring up talent."

On 21 December 2008, Pete Siebold flew WhiteKnightTwo for the first flight. White Knight had made twenty-three flights before carrying SpaceShipOne the first time. WhiteKnightTwo flew a similar number of flights, twenty-four, before lifting SpaceShipTwo on its first flight.

WhiteKnightTwo was built much bigger than required to simply carry and launch SpaceShipTwo. The payload weight is about 30 percent more than needed for SpaceShipTwo. This workhorse can haul a payload of thirty-five thousand pounds up to fifty thousand feet over a range of 2,300 miles. This gives it the capability to possibly launch small payloads into low Earth orbit, conduct scientific experimentation, and even function as a transport or utility aircraft in nonspace-related applications. Launching a single person into orbit from WhiteKnightTwo is even a "theoretical" possibility. Though WhiteKnightTwo can be used as a launch platform for different applications, Virgin Galactic's primary focus is on flying passengers to suborbital space in SpaceShipTwo.

WhiteKnightTwo made its first flight on 21 December 2008, piloted by Pete Siebold with Clint Nichols as copilot. In 2009, it made its debut at Oshkosh, as shown here. Pilots fly from the right fuselage. The left fuselage of the prototype does not have windows, but black decals are used to represent how the subsequent vehicles will look with actual windows in place. Just like White Knight before it, WhiteKnightTwo will be able to act as a trainer for SpaceShipTwo pilots. By using four spoilers to kill lift, two inboard and one outboard on each wing, WhiteKnightTwo can mimic the glide performance of SpaceShipTwo. These spoilers also eliminate the need for thrust reversers on the engines. *Dan Linehan*

WhiteKnightTwo designer Bob Morgan (center) and WhiteKnightTwo pilot Pete Siebold (right) prepare Richard Branson for his first flight in WhiteKnightTwo during the EAA AirVenture on 28 July 2009. The cabin of WhiteKnightTwo will be nearly identical to that of SpaceShipTwo. And the main difference for instruments in the cockpit is the center console where WhiteKnightTwo has four throttles and turbofan engine controls and SpaceShipTwo has rocket engine switches and levers for its feather. *Virgin Galactic/Mark Greenberg*

Rocket Engine Accident

A fatal explosion occurred as Scaled Composites conducted tests for SpaceShipTwo's hybrid rocket during the very early stage of development.

Eric Blackwell, Todd Ivens, and Glen May lost their lives on 26 July 2007. Three others were severely injured but have since recovered. Scaled Composites is a tight group because of its family-like culture, somewhat remote location, and the fact that employees can't openly discuss the sensitive projects they work on with others. The losses and injuries to these men hit the company terribly hard.

SpaceShipTwo required a much bigger rocket engine than SpaceShipOne did, but they both would use the same oxidizer. As with SpaceShipOne's rocket engine development program, early testing was done cold flow, which is like taking all the steps to fire the rocket engine except that the oxidizer and fuel are never ignited.

When the oxidizer flows from the tank during a cold flow test, it runs straight through the rest of the rocket engine without reacting with the fuel. A special igniter must be used to start the rocket engine burning. Without it, there is no combustion and fiery plume. Cold flows were considered safer and lower risk tests because flowing the oxidizer was only a physical transfer, not the chemical reaction that librated enough energy to move a spacecraft to space.

Work halted for about a year not only on the rocket engine but on SpaceShipTwo as well until internal and external investigations got to the bottom of what caused the accident.

An official report about the accident was given by the Occupational Safety and Health Administration (OSHA), which stated: "The explosion occurred during an oxidizer tank flow test in which nitrous oxide flowed through a valve on the end of a tank being developed for a rocket motor. The test was designed to be 'cold flow' and no fuel was to be present. However, during the test the nitrous oxide ignited and exploded."

This was a very dark time for Scaled Composites. Safety had always been paramount to Burt Rutan. Coming up with aircraft with very safe flying qualities, such as the stall-resistant canard designs of the VariViggen and the VariEze and the safe engine-out twin designs of the Defiant and Boomerang, was the foundation Rutan built RAF upon, which then grew into Scaled Composites.

In the thirty-three years since RAF first opened its hangar doors in Mojave, after all the aircraft developed and flight tested, neither RAF nor Scaled Composites had had a work-related fatality. It was an unthinkable tragedy, yet it still happened.

One of the probable causes of the explosion was determined to be contamination. So Scaled Composites issued the following list of modifications and procedures it would undertake as a result of the accident:

- Conduct increased compatibility testing between N_2O and any materials that contact it in the tank and eliminate incompatible materials in the flow path
- Revise cleaning procedures to further minimize the risk of contaminants in the system
- Replace the composite liner in the N_2O tank with a metal tank liner
- Dilute N_2O vapor in the tank with nitrogen or another inert gas to decrease its volatility and/or act as a pressurant
- Design additional safety systems for the N_2O tank to minimize the danger due to tank overpressure, for example, a burst disk feature
- Increase the amount of testing during the development program
- Form an advisory board comprised of rocket industry experts for oversight
- Improve test site safety procedures
- Provide the industry with any pertinent materials compatibility data and/or testing protocols developed as it moves forward

Scaled Composites resumed development of the rocket engine and SpaceShipTwo in 2008 and has since implemented these corrective actions.

It goes without saying that the loss of life and injuries sustained by these men was tragic. It is not sufficient to say that because they happened to be pursuing something pioneering, at the forefront of technology, that this circumstance was not unexpected.

However, the reality is that there are thousands of workplace-related fatalities as well as tens of thousands motor vehicle-related fatalities every year in the United States alone. These numbers are large but go relatively unnoticed because the surrounding situations are more commonplace. Do we stop working or driving cars?

It is human nature to advance, to progress, to stretch boundaries. But we have an obligation to be safety-minded and to be extremely careful not to let bottom-line decisions cloud this. However, as the commercial space industry grows, it is naive to think that it will be the only industry impervious to accidents.

Charles Lindbergh wrote this after his courageous flight in the *Spirit of St. Louis*: "I don't believe in taking foolish chances, but nothing can be accomplished without taking any chance at all."

On 26 July 2007, a rocket engine explosion claimed the lives of three Scaled Composites employees and severely injured three others. Scaled Composites had never faced a tragedy like this before. A memorial is dedicated to the three men in Legacy Park at Mojave Air & Space Port. *Dan Linehan*

A Stellar Enterprise

California Governor Arnold Schwarzenegger and New Mexico Governor Bill Richardson christened the SpaceShipTwo prototype as the *Virgin Space Ship (VSS) Enterprise*, tail number N339SS, during its rollout on 7 December 2009. *Enterprise* was chosen out of respect for the crafts that have borne this name before and the private enterprise that initiated commercial space travel.

In comparing SpaceShipOne to SpaceShipTwo, the lineage is clear in terms of mission profile, air launch, flight control, rocket engine, and feather system. But it was very important for Virgin Galactic to expand the experience its customers would have as opposed to the experience of being strapped in the backseat of a duplicate of SpaceShipOne.

As Virgin Galactic sought to figure this out, it determined there were four things that people wanted most in a journey to space: the view of Earth and the black sky of space, the thrill of a rocket ride, the experience of weightlessness, and the ability to float around in the cabin. So the design of SpaceShipTwo focused not just on achieving but on maximizing these experiences for the passengers.

SpaceShipTwo is flown by a pilot and copilot. The "shirt-sleeve" environment inside SpaceShipTwo will not require the passengers or crew to wear spacesuits. The functionality of the cockpit was also a consideration, and Virgin Galactic brought in airline pilots to help review the human factors since the vehicle is manually flown over a wide range of flight conditions and is intended to fly several times a week.

Although the overall length doubled, there is a huge increase in cabin volume by comparison. For SpaceShipOne, the seating for the pilot and two passengers was confined to small section with a 5-foot diameter that quickly narrowed down to its pointed nose. The pilot and copilot of SpaceShipTwo sit side-by-side in the flight deck, forward and slightly above the cylindrically shaped, 12-foot-long, 7.5-foot-diameter, six-passenger cabin.

One of the most noticeable differences on SpaceShipTwo is the position of the wing. On SpaceShipOne's first Ansari X Prize attempt, it started to roll uncontrollably and spiral its way up to space. This had to do with the high-wing configuration of

Scaled Composites designed a much more advanced simulator for SpaceShipTwo and WhiteKnightTwo. A giant curved projection screen outside its windows gives depth perception, so when flown the view looks very realistic. And when SpaceShipTwo pirouettes or rolls in a space simulation, it feels so real that it's hard not to grab the edges of the seat to stop from falling over. The simulator also gives a realistic feeling of the forces on the flight controls. However, Scaled Composites has not yet figured out how to simulate weightlessness. There's no substitute for the real thing. *Virgin Galactic*

When the rocket engine cuts off, SpaceShipTwo coasts to space and little by little gravity slows it down. Passengers get to unstrap from their seats and float around in the cabin, experiencing approximately 4 minutes of weightlessness and periods of microgravity as well. *Virgin Galactic*

SpaceShipOne had room for one pilot up front in the nose and two passengers in a row right behind the pilot's seat. SpaceShipTwo flies with a pilot and copilot, who sit side-by-side in a slightly raised flight deck. And SpaceShipTwo carries six passengers in its 12-foot-long and 7.5-foot-diameter cabin. *Virgin Galactic*

SpaceShipOne. This problem was solved by moving the wing down the fuselage to a low-wing configuration.

SpaceShipOne had sixteen 9-inch-diameter windows, but to afford the best sights, SpaceShipTwo has a 17-inch-diameter side window and a 13-inch-diameter overhead window for each passenger. The crew has four 21-inch-wide triangular windows and a fifth window smaller in size.

Passengers can't see a blanket of stars in suborbit on the daylight side of Earth because of all the sunlight that reflects off Earth. This is the black sky. The stars are outshined. But on the dark side of Earth where the Sun is hidden, if spaceflights operate during nighttime, then there are stars galore to see. There also is no atmosphere between the passengers and the stars to cause the stars to twinkle—just the passengers, the vacuum of space, and the stars.

Riding a Rocketplane

Once clear of the mothership, SpaceShipTwo's pilot ignites its rocket engine and in eight seconds, the spacecraft breaks the sound barrier. In about forty-five seconds, SpaceShipTwo travels faster than three times the speed of sound (Mach 3). Passengers experience a max g-force of 3.8 g on boost, but during reentry they face 6 g. They will experience approximately four minutes of weightlessness.

Virgin Galactic states it intends to fly SpaceShipTwo to "an apogee of at least 110 kilometers." This is equal to 360,900 feet or 68.4 miles. By the way of comparison, Brian Binnie flew higher than this in SpaceShipOne (112 kilometers, 367,500 feet, or 69.6 miles). Given Burt Rutan and Richard Branson's penchants for pushing boundaries, it doesn't seem reasonable the SpaceShipTwo will not fly higher than SpaceShipOne. The 110 kilometers is almost certainly understated. After all, if the max speed is expected to be higher, and the max reentry force is higher, then it should follow that the max altitude will be higher even if the gravitational force on SpaceShipTwo is a little higher because of its greater mass. SpaceShipTwo has a design range of 70–130 kilometers after all.

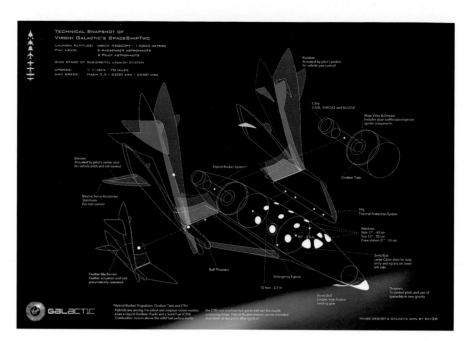

This diagram from Virgin Galactic gives some of the specifications for SpaceShipTwo. But these figures were given before the flight test program started. So slight variations might exist as the flight program and daily operation reveal SpaceShipTwo's actual capabilities. *Virgin Galactic*

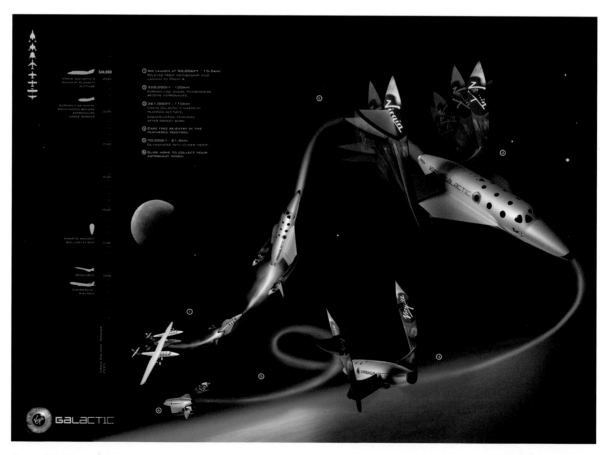

SpaceShipTwo will fly a very similar profile to SpaceShipOne. However, SpaceShipTwo will fly faster and experience higher g-forces on reentry. Apogee is planned to be greater than 110 kilometers, and the entire spaceflight, from takeoff to touchdown, is expected to last 2 hours. *Virgin Galactic*

Spaceport America

Just like the Mojave Desert, the southern desert of New Mexico claims home to many pioneering advancements in aerospace. Putting aside what *actually* happened in Roswell, the United States space program took foothold when the V-2 rockets captured during World War II were taken to White Sands Missile Range and tested. The V-2 rocket was the starting equipment used by both the United States and the former Soviet Union during their track meet up and down from space that started with Sputnik.

Ironically, the V-2 was coming back to its roots. Robert Goddard, an inventor and visionary, chose this region to conduct his rocket testing in the 1930s. The rocket scientists who had developed the V-2 had built upon Goddard's discoveries involving liquid-fueled rocket engines.

This part of New Mexico has been rocket country for about as long as there have been big rockets. After Virgin Galactic announced it would begin a spaceline with SpaceShipTwos and WhiteKnightTwos, New Mexico moved fast to make an offer Virgin Galactic couldn't refuse.

It wasn't just New Mexico's rocket legacy in itself that was the draw. It was very practical reasons. Spaceport America is located forty-five miles northeast of Las Cruces. It is dry, has relatively good weather year-round, and has an elevation of 4,600 feet, which is nearly a mile closer to space than at sea level. The location is somewhat remote, so there are no major population centers to contend with. And the airspace is uncongested, especially because of the restricted zones of nearby White Sands Missile Range. Las Cruces and Holloman Air Force Base even played host to the X Prize Cup in the years following the Ansari X Prize.

Spaceport America broke ground in 2009 and completed its ten-thousand-foot-long runway in 2010. In 2011, it started to become operational. The centerpiece of the spaceport is the dynamic, eye-shaped terminal hangar facility (THF). With room to hangar the SpaceShipTwos and WhiteKnightTwos of Virgin Galactic, there is also room for vehicle maintenance, astronaut training, public viewing of the runway, and more.

Spaceport America designed by URS/Foster + Partners
Conceptual image courtesy of Vyonyx Ltd

In 2005, New Mexico announced it would invest $200 million to construct Spaceport America. The spaceport received its launch license from the FAA in 2007. This is where Virgin Galactic will start commercial service to space in SpaceShipTwo. *Spaceport America Conceptual Images URS/Foster + Partners*

The THF was built into the desert floor for insulation and makes use of natural lighting, cooling, and ventilation as well as solar power to make the facility highly energy efficient.

Spaceport America is owned by the State of New Mexico and operates like an airport. But in this case, spacelines and other space companies lease the use of the spaceport. Virgin Galactic is the anchor tenant and has signed a twenty-year lease with Spaceport America and will locate its main operations base there.

Flight testing on the prototypes of SpaceShipTwo and WhiteKnightTwo and those built subsequently will be conducted in Mojave. But once they are turned over to Virgin Galactic, they will be operated commercially out of Spaceport America.

You don't have to be a billionaire or an aerospace genius to be part of this new wave of space travel. The citizens of New Mexico helped front the money to build Spaceport America. The surrounding communities even voted to increase their taxes to ensure it would be built. As a thank you to those people of who voted in favor of the tax hike, Virgin Galactic will hold a drawing to give one of these voters a free to trip to space.

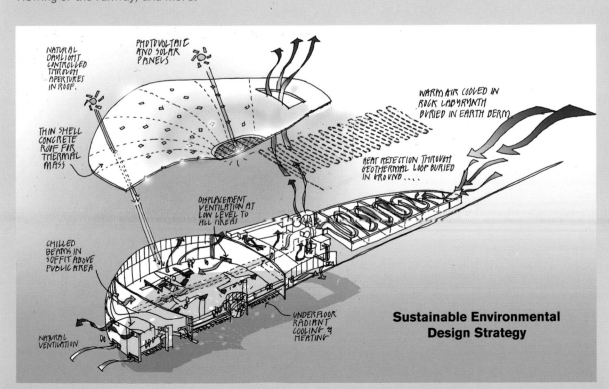

Spaceport America integrates state-of-art green building technology into a stunning and imaginative design that looks more like artwork than space work. *Spaceport America Conceptual Images URS/ Foster + Partners*

NATURAL DAYLIGHT CONTROLLED THROUGH APERTURES IN ROOF.

PHOTOVOLTAIC AND SOLAR PANELS

THIN SHELL CONCRETE ROOF FOR THERMAL MASS

WARM AIR COOLED IN ROCK LABYRINTH BURIED IN EARTH BERM

HEAT REJECTION THROUGH GEOTHERMAL LOOP BURIED IN GROUND

DISPLACEMENT VENTILATION AT LOW LEVEL TO ALL AREAS

CHILLED BEAMS IN SOFFIT ABOVE PUBLIC AREA

NATURAL VENTILATION

UNDERFLOOR RADIANT COOLING & HEATING

Sustainable Environmental Design Strategy

SpaceShipTwo's hull is shown being pieced together. These composite sections will be bonded together much the same way SpaceShipOne was. The section being added is the top half of the passenger cabin and cockpit. At the back end is the aft pressure bulkhead. *Virgin Galactic*

Rocket Engine

SpaceDev, having developed several of the key internal components on Scaled Composites' SpaceShipOne rocket engine and now a subsidiary of Sierra Nevada Corporation, rejoined the team at Scaled Composites in August 2008 to assist in the development of the SpaceShipTwo rocket engine's internal components. Like SpaceShipOne, the oxidizer is nitrous oxide and the fuel is a solid material. The synthetic rubber used as fuel by SpaceShipOne may not necessarily be the same fuel used by SpaceShipTwo. The rocket engine can also be shut down at any time after ignition by simply closing off the flow of the oxidizer.

In May 2009, the first phase of rocket engine development completed. Scaled Composites then published safety guidelines for nitrous oxide usage relating to rocket engines in June 2009.

As a result of the modifications to the rocket engine, SpaceShipTwo had to be modified as well. Two pressurized tanks of helium were added. As a safety precaution, helium from these tanks will feed into the oxidizer tank to keep it fully pressurized as oxidizer flows out of the tank and into the rocket engine. This has the added effect of making the rocket ride a little bit smoother compared to SpaceShipOne, where after about one minute of running, the oxidizer remaining in the tank makes a liquid-to-gas transition. This caused SpaceShipOne's rocket engine to chug along for a few seconds.

As more sections get added, SpaceShipTwo resembles more and more its final configuration. SpaceShipTwo has a dual hull made of carbon fiber composite outer and inner shells that sandwich a Nomex honeycomb core. *Virgin Galactic/Thierry Boccon-Gibod*

Flight Testing to Early Operation

There is also a difference between the two flight test programs. SpaceShipOne had to make two spaceflights to win the Ansari X Prize. Anything after that was gravy. However, SpaceShipTwo will be flown time and time again. And once flight testing is over, it won't be flown by test pilots anymore. It is designed to be flown by airline pilots. And where SpaceShipOne could—and did—fill its two passenger seats with four hundred pounds of ballast, that arrangement won't quite work on a commercial passenger-carrying spacecraft.

The flight-testing program for SpaceShipTwo will obliviously be used to confirm its capabilities, performance, and operation during spaceflight. But since SpaceShipTwo will be used commercially—that is, generating money from being flown like an airliner—things like turnaround time, consumable requirements, and preventative maintenance will start to be determined.

As with the SpaceShipOne program, SpaceShipTwo and WhiteKnightTwo have been designed with as many common components as possible. WhiteKnightTwo started flying more than a year earlier than SpaceShipTwo, so this gave Scaled Composites a head start on understanding how some the SpaceShipTwo components preformed in an operational environment by being able to use WhiteKnightTwo to shake things out ahead of time.

The SpaceShipTwo flight test program will follow a similar progression to that completed by SpaceShipOne. Captive carry, glide, rocket-powered, and then space test flights will incrementally expand the flight envelope. SpaceShipTwo had its first captive carry flight on 22 March 2010.

On 10 October 2010, pilot Pete Siebold and copilot Mike Alsbury flew SpaceShipTwo on its first glide flight. "Our spaceship demonstrated impressive flying qualities right out of the box. Its flight test-measured stability and gliding performance exceeded the pre-flight predictions," Rutan said in his official announcement. "The test crew opened up two-thirds of SpaceShipTwo's required subsonic speed envelope, maneuvered it above 2 g, checked its dynamic and sideslip handling, exercised its flight-path control system and made a perfect landing; spot-on the runway target."

Following a successful second glide flight on 28 October 2010, SpaceShipTwo made its third glide flight on 17 November 2010. The test flight lasted 11 minutes and 39 seconds and achieved all its objectives, including expanding SpaceShipTwo's flight envelope to 283 miles per hour and 3.5 g.

But even after the completion of the flight test program, Virgin Galactic wants continued improvement and refinement as SpaceShipTwo and WhiteKnightTwo mature and the knowledge base grows. This includes optimizing performance and making the manufacturing process more efficient.

Burt Rutan sits here in the very roomy inside of SpaceShipTwo during the very early stages of construction. In fact, at this stage it is hard to tell the inside of SpaceShipTwo from the inside of WhiteKnightTwo. The fuselage of SpaceShipTwo was designed to be the same inner and outer shapes as the two booms of WhiteKnightTwo. *Virgin Galactic*

Though modified with additional safety features, SpaceShipTwo will use a hybrid rocket engine with a cantilever design as did SpaceShipOne. It will also use liquefied laughing gas (N_2O or nitrous oxide) as the oxidizer and likely solid synthetic rubber (HTPB or hydroxyl-terminated polybutadiene) as the fuel. In this test firing, the rocket engine uses the same components and systems that will be used in SpaceShipTwo. *Virgin Galactic*

As the SpaceShipTwo and WhiteKnightTwo fleet grows, The Spaceship Company will individually flight test each vehicle. This procedure is different than for a certified airliner, where once the type is certified, every new aircraft built of that type doesn't go through a flight test program.

As with airliners, the Federal Aviation Administration (FAA) has set up regulations for the operation of suborbital spacecraft. But a much larger hurdle that Scaled Composites and Virgin Galactic have faced is the International Traffic in Arms Regulations (ITAR). If this sounds like a strange thing to have to deal with, just remember that suborbital technology is after all the basis behind ballistic missiles. So as the first step, Virgin Galactic will satisfy ITAR requirements by operating solely in the United States. To fly outside the United States would require different clearances and will be part of Virgin Galactic's future plans.

Even as far as the operation of SpaceShipTwo goes, since Virgin Galactic is really at the start of a whole new industry, it will be evolving as it learns from flying back and forth to space time after time. As Virgin Galactic demonstrates forward progress, it can expand in ways

SpaceShipTwo is more stretched out in relation to SpaceShipOne. This has the effect of moving the feathered tail booms farther back in proportion to the length of the fuselage. SpaceShipOne's feather pivots about halfway from the tip of the nose, and SpaceShipTwo's pivots about two thirds of the way back. This is obviously an advantage for SpaceShipTwo. While passengers certainly want the feather up, not many passengers would want a window with an obstructed view. *Dan Linehan*

Flight testing for SpaceShipTwo began with captive carries. Kind of like using training wheels on a bicycle, once Scaled Composites felt comfortable with SpaceShipTwo's flight characteristics with a little help from WhiteKnightTwo, the next step was to take off the training wheels and let SpaceShipTwo fly free. Glide flights followed by rocket-powered flights and spaceflights would be the same envelope expansion progression that had been done with SpaceShipOne. *Virgin Galactic/Mark Greenberg*

SpaceShipTwo made its maiden flight on 10 October 2010 piloted by Pete Siebold with copilot Mike Alsbury. SpaceShipTwo released from WhiteKnightTwo at 46,000 feet and expanded the flight envelope out to 207 miles per hour and 2 g. During its 13 minutes of flying time, SpaceShipTwo completed all its objectives and preformed better than expectations. *Virgin Galactic/Mark Greenberg*

that have been planned for and ways that could be completely unexpected—whether it is variations in the flight profile, the frequency launch, the training, the ground experience, or the launch and landing sites. Time and space will tell.

A Ticket to Ride

Included in the two hundred thousand-dollar price tag of the trip to space is a three-day ground experience where passengers get training on their route to earning astronaut wings. The training covers familiarization with SpaceShipTwo and spaceflight both on the ground and aboard WhiteKnightTwo. Since WhiteKnightTwo is designed to be a trainer as well, passengers will get to experience some of the low g-force and high g-force feelings they will encounter during their ride on SpaceShipTwo. Part of the preparation is also devoted to team building with the other passengers who will fly to space together. The training is ultimately designed by Virgin Galactic to help maximize the passengers' enjoyment when it comes time to flick the rocket ignition switch.

By September 2010, Virgin Galactic already had 370 future astronauts signed up, who put just more than $50 million down in deposits. The first hundred passengers, called Founders by Virgin Galactic, will have

the order in which they fly to space chosen by a random drawing. These tickets are already sold out. Pioneers will make up passengers 101 to 500 and then Voyagers after that. For these next two groups, a passenger's order in the space cue depends upon when they made their reservations and the size of their space deposit.

Virgin Galactic envisions flying fifty thousand passengers to space in the first ten years of operation. Fortunately for most, passengers don't have to be of Mercury Seven material. Virgin Galactic reports, "Early indicators show that the required medical assessment will be simple and unrestrictive and that the vast majority of people who want to fly will not be prevented from doing so by health or fitness considerations."

As Virgin Galactic moves forward, rides through the aurora borealis, rides during sunset and during sunrise, and all kinds of possibilities await.

And to help facilitate booking a spaceflight aboard SpaceShipTwo, Virgin Galactic has coordinated a global network of accredited space travel agents. But like any new technology that hits the market, its price is high. This has to do partly with recovering development costs. But as the infrastructure matures, the economy of scale increases, and competition enters the game, the cost of a ticket to space will become more down to Earth, but the ride will still be out of this world.

Afterword: A Star to Steer By

Shown circling the Moon in his VariViggen, Burt Rutan looks forward to the day when he can board a space hotel and then launch out of Earth orbit to swing around the Moon. And he has just the design in mind to do it. *Courtesy of Burt Rutan*

The days of Apollo are long gone. It was a different world back then. The days of Space Shuttle are gone, too. It is a different world now. I was too young for Apollo to know any better, but I did grow up with a copy of a Space Shuttle manual, books about the F-14 Tomcat, instrument panel posters of the aircraft my dad flew for the airlines, and my imagination.

Some of the most influential people in the aviation industry were kids around the time of the technological burst in aviation that occurred shortly following the Wright brothers' first flight. And a lot of the pioneers of new space were kids during Apollo. But the question is, what do the young minds of today have to look forward to? What will they be watching?

In 2008 something extraordinary happened to me. My book *SpaceShipOne: An Illustrated History* had just come out, and I was at EAA's AirVenture in Oshkosh sitting among two rows of other authors in a giant warehouse filled with books. A little girl walked in and came straight over to me without even glancing at the shelves of books or other authors. Before I could even say hello to her, she said this: "I am five years old. And by the time I am ten, I'll be on the Moon. My parents said I could have any book I want. And I want yours."

Holy moly, right? So I asked her if I could sign it for her. She agreed. I wrote that I hoped she would become a great spacepilot and visit the Moon and stars. And if she did, would she give me a ride?

I handed her the book. She read the inscription and gave an emphatic, "Yes."

The next day, at the same place, she came back and marched right up to me again and said, "I read the first chapter. And it was very good." And then she marched off.

I certainly did not write that book for five-year-olds. And I wonder how much she has progressed over the years. I wonder how many others like her are out there with parents who read to them and encourage their imaginations and learning.

I wonder who will be the next Burt Rutan aiming for the stars—homemade models to homebuilt aircraft to hometown spaceships. Will she be the one?

How long will it take Virgin Galactic and all the other innovators in this new space age for their visions to turn to reality? Is this now where the seeds are planted for the next generation of pioneers?

We borrow this planet for such a short time. And as individuals we want to make a better world for our offspring. But this world has become very interconnected and complicated, and this job has become much more difficult.

I cannot wait for the time when fifty thousand people make it to space and get to see this beautiful blue planet. I cannot wait for each and every one of them to take this overview and return with it back to Earth.

Glossary of Acronyms, Abbreviations, and Conversion

AFB	air force base
ATTT	Advanced Technology Tactical Transport
cc	cubic centimeter
CG	center of gravity
CRV	crew return vehicle
CTN	case/throat/nozzle
DARPA	Defense Advanced Projects Research Agency
EAA	Experimental Aircraft Association
ERAST	Environmental Research Aircraft and Sensor Technology
FDD	flight director display
ft	feet (1.00 foot = 0.305 meters)
ft²	square feet
fpm	feet per minute
g	acceleration of gravity
gal	gallon (1.00 gallon = 3.79 liters)
hp	horsepower
HTPB	hydroxyl-terminated polybutadiene (synthetic rubber)
KEAS	knots equivalent airspeed
KISS	keep it simple stupid
kt	knot, nautical mile per hour (1.00 knot = 1.15 mile per hour)
lbs	pounds (1.00 pound = 0.454 kilograms)
LEZ	Long-EZ
L/D	ratio of lift to drag
LOX	liquid oxygen
mi	miles (1.00 mile = 1.61 kilometers)
MONODS	mobile nitrous oxide delivery system
mph	miles per hour
NGBA	Next Generation Business Aircraft
NGT	Next Generation Trainer
nm	nautical mile (1.00 nautical mile = 1.15 statute mile)
N₂O	nitrous oxide (laughing gas)
O₂	oxygen (molecular)
POC	proof of concept
RAF	Rutan Aircraft Factory
R&D	research and development
RC	radio control
RCS	radar cross section or reaction control system
RLV	reusable launch vehicle
RPV	remotely piloted vehicle
RTW	round-the-world
SCAT	Scaled Composites Advanced Turboprop
SCUM	Scaled Composites unit mobile
sm	statute mile
SME	shuttle main engine
SMUT	Special Mission Utility Transport
SRB	solid rocket booster
SS1	SpaceShipOne
SS2	SpaceShipTwo
STOL	short takeoff and landing
THF	Terminal Hangar Facility
TONU	Tier One navigation unit
TPS	thermal protection system
TSC	The Spaceship Company
TST	test stand trailer
VTOL	vertical takeoff and landing
UAV	unmanned aerial vehicle
USAF	United States Air Force
VMS	Virgin Mother Ship
VSS	Virgin Space Ship
WK	White Knight
WK2	WhiteKnightTwo
wt	weight

Appendix: Partial List of Rutan Aircraft Factory and Scaled Composites Models

Model	Name	Customer	Fabrication	First Flight
1	A-12	-	-	-
27	VariViggen	homebuilders	Rutan Aircraft Factory	Rutan Aircraft Factory
28	MiniViggen	-	-	-
31	VariEze POC	R&D	Rutan Aircraft Factory	Rutan Aircraft Factory
32-SP	VariViggen SP	homebuilders	Rutan Aircraft Factory	Rutan Aircraft Factory
33	VariEze	homebuilders	Rutan Aircraft Factory	Rutan Aircraft Factory
35	AD-1	NASA	Ames Industrial	NASA
40	Defiant POC	R&D	Rutan Aircraft Factory	Rutan Aircraft Factory
45	-	-	-	-
49	-	-	-	-
54	Quickie	Quickie Airplane Corp.	Rutan Aircraft Factory	Rutan Aircraft Factory
59	-	-	-	-
61	Long-EZ	homebuilders	Rutan Aircraft Factory	Rutan Aircraft Factory
-	Power-Augmented-Ram Landing Craft (PARLC)	U.S. Navy	Ames Industrial	-
61-B	Jet LEZ Vantage	proprietary	Scaled Composites	Scaled Composites
61-PD	Borealis	Air Force Research Laboratories/Innovative Science Solutions	Scaled Composites	Scaled Composites
61-R	Rodie	proprietary	Scaled Composites	Scaled Composites
68	AMSOIL Biplane Racer	Danny Mortensen/AMSOIL	customer	customer
69	-	-	-	-
72	Grizzly	R&D	Rutan Aircraft Factory	Rutan Aircraft Factory
73	Next Generation Trainer (NGT)	Fairchild Republic	Ames Industrial	Rutan Aircraft Factory
74	Defiant	homebuilders	Fred Keller	Rutan Aircraft Factory
76	Voyager	Voyager Aircraft	Rutan Aircraft Factory	RAF/Voyager Aircraft
77	Solitaire	homebuilders	Rutan Aircraft Factory	Rutan Aircraft Factory
78-1	Commuter	-	-	-
81	Catbird	Beechcraft	RAF/Scaled Composites	Scaled Composites
97	Microlight	Colin Chapman/Lotus	Scaled Composites	Scaled Composites
111	-	-	-	-
112	-	-	-	-
113	-	-	-	-
115	Starship POC	Beechcraft	Scaled Composites	Scaled Composites
120	Predator	Advanced Technology Aircraft Corp.	Scaled Composites	Scaled Composites
127	-	-	-	-
-	Scarab	Teledyne Ryan Aeronautical	Scaled Composites	-
133	Advanced Technology Tactical Transport (ATTT)	DARPA	Scaled Composites	Scaled Composites
-	Stars and Stripes Wing Sail (H1 and H2)	Sail America	Scaled Composites	-
133-B	ATTT Bronco Tail	DARPA	Scaled Composites	Scaled Composites
-	Searcher	Israeli Aircraft Industries	Scaled Composites	-
143	Triumph	Beechcraft	Scaled Composites	Scaled Composites
144	CM-44	California Microwave	Scaled Composites	Scaled Composites
151	Agile Responsive Effective Support (ARES)	R&D	Scaled Composites	Scaled Composites
-	Lima 1	Toyota	Scaled Composites	Scaled Composites
158	Pond Racer	Bob Pond	Scaled Composites	customer
-	Su-25 Rocket-on-a-Rope (ROAR)	Sandia National Laboratories	Scaled Composites	-
173	TFV	Loral	Scaled Composites	-
175	B-2 RCS	Northrop Corp.	Scaled Composites	-
179	PLADS/Rockbox	Lockheed	Scaled Composites	-
-	Pegasus	Orbital Sciences Corporation	Scaled Composites	-
181	Earthwinds	Richard Branson/Earthwinds	Scaled Composites	customer
191	Lima 2	Toyota	Scaled Composites	Scaled Composites
-	Ultralite	General Motors	Scaled Composites	-
202	Boomerang	Rutan Designs	Rutan Designs	Rutan Designs
226	Raptor D-1	U.S. Department of Energy	Scaled Composites	Scaled Composites
-	Z-40 Bladerunner	Zond	Scaled Composites	-
226-B	Raptor D-2	U.S. Department of Energy/NASA	Scaled Composites	Scaled Composites
231	Eagle Eye	Bell Helicopter	Scaled Composites	-
-	DC-X	McDonnell Douglas	Scaled Composites	-
233	Freewing	Freewing Aircraft	Scaled Composites	-
-	Comet	Space Industries	Scaled Composites	-
-	Kistler Zero	Kistler Aerospace	Scaled Composites	-
247	Vantage	VisionAire Corp.	Scaled Composites	Scaled Composites
257	Global Hilton	Voyager Aircraft	Scaled Composites	Voyager Aircraft
267	-	-	-	-
271	V-Jet II	Williams International	Scaled Composites	Scaled Composites
276	X-38	NASA	Scaled Composites	-
-	Fuji Mini Shuttle POC	-	-	-
280	-	-	-	-
281	Proteus	Wyman-Gordon	Scaled Composites	Scaled Composites
-	Roton	Rotary Rocket Company	Scaled Composites	Rotary Rocket Company
-	Nosejob	-	-	-
287	Alliance	NASA	Scaled Composites	-
302	TAA-1	Toyota	Scaled Composites	Scaled Composites
-	Seeker	International Systems	-	-
309	Adam 309	Adam Aircraft	Scaled Composites	Scaled Composites
311	Capricorn GlobalFlyer	Steve Fossett/Virgin Atlantic	Scaled Composites	Scaled Composites
312	-	-	-	-
313	-	-	-	-
316	SpaceShipOne	Mojave Aerospace Ventures	Scaled Composites	Scaled Composites
317	-	-	-	-
318	White Knight	Mojave Aerospace Ventures	Scaled Composites	Scaled Composites
326	X-47	Northrop Grumman	Scaled Composites	-
339	SpaceShipTwo	Virgin Galactic	Scaled Composites	Scaled Composites
348	WhiteKnightTwo	Virgin Galactic	Scaled Composites	Scaled Composites
356	-	-	-	-

First Flight Date	Tail Number	Description
-	-	RC model of single- place, canard pusher
18 May 1972	N27VV	canard pusher, two-place, wood construction, single-engine, plans ready February 1974
-	-	canard pusher, two-place, based off BD-5
21 May 1975	N7EZ	canard pusher, two-place, high-efficiency, Volkswagen engine
July 1975	N27VV	Model 27 with higher aspect ratio composite wings, plans ready November 1975
14 March 1976	N4EZ	improved version of Model 31, aircraft engine, plans ready July 1976
20 December 1979	N805NA	skew wing research aircraft, single-place, twin turbojets
30 June 1978	N78RA	tandem-wing, four-place, push-pull engines
-	-	single-engine, four-place, small airplane
-	-	early concept for Model 54, like a single-place miniature Model 33, not built
17 November 1977	N54Q	tandem-wing, single-place, Onan engine
-	-	biplane with joined wingtips, early concept of Model 120, not built
13 June 1979	N79RA	canard pusher, two-place, high-efficiency, long-range, single-engine, plans ready April 1980
September 1980	-	twin turbojet engines at bow put vessel into ground effect to reduce drag
August 1993	N3142B	Model 61 powered by FJ107 jet engine
31 January 2008	N90EZ	pulse detonation engine research aircraft using Model 61
August 2001	N41GB	based off Model 61
August 1981	N301LS	tandem-wing, single-place, biplane class air racer, single-engine
-	-	concept similar to Model 68 but with joined wingtips, not built
22 January 1982	N80RA	three-surface, four-place, STOL, bush plane, single-engine
10 September 1981	N73RA	subscale flight demonstrator for proposed T-46 trainer, twin turbojets
16 July 1983	N39199	improved version of Model 40, plans ready June 1984
22 June 1984	N269VA	trimaran, two-place, designed to fly around the world nonstop and nonrefueled, push-pull engines
28 May 1982	N81RA	canard, single-place, self-launching sailplane, plans ready August 1983
-	-	model of 36-passenger aircraft, canard, tail-mounted push-pull engines
14 January 1988	N187RA	three-surface, five-place, very low drag, turbocharged engine
30 January 1983	N97ML	canard pusher, two-place, rigid ultralight, single-engine
-	-	early concept for ARES, joined wing
-	-	early concept for Starship
-	-	early concept for Starship
29 August 1983	N2000S	variable geometry canard, twin turboprop pushers, subscale flight demonstrator for next generation business aircraft
17 September 1984	N480AG	three-surface, single-place, crop duster, single-engine
-	-	long-range, high endurance vehicle
June 1986	-	reconnaissance drone, solid rocket engine and turbojet engine
29 December 1987	N133SC	three surfaces, STOL, subscale flight demonstrator, twin turboprop engines
May 1988	-	carbon fiber wing sails for America's Cup Race, 90 ft and 108 ft spans
December 1988	N133SC	Model 133 with reconfigured tail
December 1988	-	pusher engine, H-tail, UAV
12 July 1988	N143SC	three-surface, seven-place, business aircraft, twin turbofan FJ-44 engines
27 February 1987	N935SC	canard pusher, manned/unmanned, long-endurance, turbocharged engine
19 February 1990	N151SC	asymmetric fuselage, canard, single-place, close air support combat aircraft, JT15D turbofan engine
April 1990	N178AE	testbed for Lexus engine
22 March 1991	N221BP	trimaran, single-place, unlimited class air racer, two 1,000-hp Electramotive engines
April 1991	-	subscale rocket-powered, cable-mounted decoy
July 1989	-	towed decoy vehicle
-	-	pole model subscale model for RCS testing
November 1989	-	parachute delivery vehicle for eight personnel
April 1990	-	air launched rocket for delivering payloads into space
November 1991	-	pressurized gondola for round-the-world balloon flight
4 November 1991	N191SC	high aspect ratio wing, conventional tail, multipassenger, single tractor engine
January 1992	-	concept car, four passengers, four doors, high efficiency
19 June 1996	N24BT	asymmetrical configuration, five-place, safe engine out qualities, long-range, twin engines
9 May 1993	N62270	high-altitude UAV for boost phase missile intercept
April 1994	-	blades for wind turbine
24 August 1994	N2272C	high-altitude UAV for Environmental Research Aircraft and Sensor Technology program (ERAST)
June 1993	-	tilt-rotor UAV
August 1993	-	subscale flight demonstrator, single-stage rocket technology
October 1994	-	tilt-wing UAV
1995	-	reentry capsule
-	-	two-stage demonstration rocket
14 November 1996	N247VA	high aspect ratio wing, conventional tail, seven-place, business aircraft, JT15D turbofan engine
January 1998	-	pressurized gondola for round-the-world balloon flight
-	-	early concept for SpaceShipOne, February 1995
13 April 1997	N222FJ	forward-swept wing, V-tail, five-place, two turbofan jet engines
12 March 1998	-	crew rescue vehicle for International Space Station
-	-	
-	-	early concept for SpaceShipOne, April 1996
26 July 1998	N281PR	tandem-wing, two-place, high-altitude, utility, two turbofan jet engines
March 1999	N990RR	space launch vehicle, rotor system used for landing
-	-	
-	-	RC model for Environmental Research Aircraft and Sensor Technology program (ERAST)
31 May 2002	N72TA	low-wing, conventional tail, four-place, fixed landing gear, IO-360 engine
2002	-	pusher engine, H-tail, UAV
21 March 2000	N309A	bronco tail, six-place, business/personal aircraft, push-pull, twin turbocharged engines
5 March 2004	N277SF	trimaran, single-place, designed to fly around the world nonstop and nonrefueled, single turbofan engine
-	-	early concept for SpaceShipOne, September 1999
-	-	early concept for SpaceShipOne, October 1999
7 August 2003	N328KF	suborbital spacecraft, three-place, high wing, feather reentry, hybrid rocket engine, February 2000
-	-	VTOL light aircraft, tail sitter
1 August 2002	N318SL	tirmaran, three-place, high-altitude, airborne launch aircraft, twin jet engines
27 July 2001	-	combat UAV, flying wing, no tail, turbofan jet engine
10 October 2010	N339SS	suborbital spacecraft, eight-place, low-wing, feather reentry, hybrid rocket engine
21 December 2008	N348MS	twin boom, sixteen- place, high-altitude, airborne-launch aircraft, four jet engines
-	-	eight-place version of Model 202

Index